"Grounded in solid research, this book is heartfelt,
practical suggestions and effective tools. Sharp, i
Tara Cousineau shows us how live from a strong h

—**Rick Hanson, PhD**, author of

"A lovely book, replete with simple kindness and full of reminders of how to enjoy and embody a kindful life."

—**Jack Kornfield**, author of *The Wise Heart*

"This is a wonderful book that focuses on kindness, bringing in work from a variety of different disciplines. If you want to learn to bring more kindness into your life, this book is a good place to start."

—**Kristin Neff**, author of *Self-Compassion*, and associate professor in the department of educational psychology at The University of Texas at Austin

"Using practical techniques supported by research, Tara offers simple yet powerful exercises that reconnect us to qualities of compassion and kindness, and in the process, incline our minds and hearts to dwell in these states more naturally. *The Kindness Cure* gives us carefully considered, warmly delivered keys to a deeper experience of kindness and connectedness."

—**Sharon Salzberg**, author of *Real Happiness* and *Real Love*

"If there were a single practice capable of transforming person and planet, it would be kindness. If you were to read a single book on kindness, let it be this one. Blending her expertise in psychology with her deep life experiences, Tara Cousineau offers you a path for cultivating kindness within and without."

—**Rabbi Rami Shapiro**, author of *The Sacred Art of Lovingkindness*

"Kindness truly connects people across their differences, and is the key ingredient in social change which this book beautifully demonstrates. Through a research lens, Cousineau takes us on a journey that is witty, relatable, and encouraging. It's a practical guide to understanding just how much kindness can reshape our world for the better. A must-read.'"

—**Jaclyn Lindsey**, cofounder and CEO of www.kindness.org

"For many years, Tara Cousineau has shared her message about the power of kindness and compassion to lift others out of the distress and despair of everyday living: from daily hassles to wrestling with inner feelings of low self-worth, and sense of disconnection. Putting her message into a book about our innate ability to care and be kind, and how to nurture this capacity, is timely and much needed. I, for one, would welcome it as a resource for clients and colleagues."

—**Nancy Etcoff, PhD**, author of *Survival of the Prettiest*, and assistant clinical professor and psychologist at Harvard Medical School and the Massachusetts General Hospital Department of Psychiatry

"Our world is hungry for kindness, and I know Tara Cousineau as someone who is dedicated to inviting more kindness, compassion, and understanding through her writing. I highly recommend *The Kindness Cure* for the wisdom it contains."

—**Donald Altman**, author of *101 Mindful Ways to Build Resilience, The Mindfulness Toolbox*, and *One-Minute Mindfulness*

"We have chased personal success and happiness to the point of exhaustion. No matter how much we achieve or acquire, we come up empty, lonely, bereft. In short, we are in need of a cure. But, as Tara Cousineau so elegantly and thoroughly reminds us, our medicine is in our DNA. We are wired for caring, connection, and compassion. Read *The Kindness Cure* and pass it on. Let the 'infection of kindness' begin!"

—**Janet Conner**, author of *Writing Down Your Soul*, *The Lotus and The Lily*, *Find Your Soul's Purpose*, and more

"In her warm and wise voice, Tara Cousineau reminds us that beneath our often frenzied and distracted lives, kindness is at the heart of our being. Even with the practice of mindfulness, life is difficult for everyone. Kindness is an action we can perform in almost every moment; it connects us to others and brightens all our lives."

—**Susan M. Pollak**, coauthor of *Sitting Together*, and president of The Institute for Meditation and Psychotherapy

"Tara Cousineau has created a groundbreaking, beautifully written book on kindness. It is a forgotten concept, but one which is vital for our health and well-being, especially in the times we live in now."

—**Alice Domar, PhD**, coauthor of *Live a Little!* and *Self-Nurture*

the kindness cure

How *the* Science *of* Compassion Can Heal Your Heart & Your World

TARA COUSINEAU, PhD

New Harbinger Publications, Inc.

Publisher's Note

This publication is designed to provide accurate and authoritative information in regard to the subject matter covered. It is sold with the understanding that the publisher is not engaged in rendering psychological, financial, legal, or other professional services. If expert assistance or counseling is needed, the services of a competent professional should be sought.

Distributed in Canada by Raincoast Books

New Harbinger Publications, Inc.
5674 Shattuck Avenue
Oakland, CA 94609
www.newharbinger.com

Cover design by Amy Shoup

Illustrations by Pamela Best

Acquired by Jess O'Brien

Edited by Marisa Solis

Library of Congress Cataloging-in-Publication Data on file

20 19 18

10 9 8 7 6 5 4 3 2 1 First Printing

For Sophie and Josie

Contents

PART 3: Kindfulness

PART 4: Kindsight

Foreword

The Kindness Cure is a peerless book on kindness that exceeds any existing work on the subject. Clinical psychologist Tara Cousineau, PhD, is an advocate for a deeper sense of our shared humanity and a practitioner of secular spirituality, with its mindfulness and practical techniques intended to elevate kindness. If you want to be a little kinder to people, including yourself, this book points you in the right direction. It combines wisdom, great stories, spirituality, useful tips, psychological insight, and good science with clear writing that anyone can enjoy as the words—and the many wonderful sketched images—jump right off the page and into the mind and soul of the reader. Rarely has a book been better able to draw the reader into an experiment with the potential for personal transformation and growth.

Why kindness in particular? Because "kindness" is an everyday word for everyone. It is a simple concept, and easy to understand. It is based on no special revelation, no esoteric truths, nor any rare insight into ultimate reality. Because "kindness" is a humble word, it does not invite arrogant debates and conflicts that pit *us* against *them*. It is refreshingly unifying; we all know what it means when someone asks if we could have been a little kinder.

Tara uses the word "kindness" because we all know it when we experience it. A lens on kindness focuses the mind on the small details of our day-to-day lives. These details are palpable: facial expression, tone of voice, attention to details, a listening pause, humility that never crowds others out of the room, or even just a simple "thank you." These are not big things, but they are the basic ingredients of a life well lived. Kindness reveals the dignity, value, and beauty of all the people we encounter on the path of everyday life. It is not an abstract "love for humanity," but a concrete concern, a local way of engaging the world right where you are. We are all the right person in the right place at the right time to be kind, and Tara helps us see that the effort is worth it.

The word "kindness" also conveys a sense of equal regard, a feeling for oneness and connectivity in the web of life. Tara defines kindness as love in action. It is the love that finds the happiness, security, and well-being of another to be as meaningful and real as our own. This definition of kindness as a manifestation of love touches on the deepest dimensions of the heart and mind, where each of us is recognized and felt to be of profound, inherent, and equal value. There can be no meaningful or valid expression of love without the modality of kindness, for it is impossible to imagine love without kindness—and without some laughter and loyalty as well. In kindness, there is no narcissism, no crowding others out, but rather an open door for everyone—equally valued and accepted, despite our imperfections.

Kindness is more important now than ever before because it refuses the distinctions that separate us: class, race, religion, ethnicity, education, gender, poverty, sexual identity, disability, and background. Kindness affirms each person, without exception, in the here and now. It need not be naïve nor unwise, but it must always see the goodness within and strive to reveal it.

As a medical professional who has spent considerable time around dying people, I can attest that the greatest regret expressed is often, "I should have been kinder." But hopefully, with Tara's help, we can all focus on kindness long before we speak our last words. Ultimately, this lovely book is also a huge challenge. In addition to changing our outward behavior, it prompts us to look inward, asking ourselves: *Am I missing out on kindness? Have I lost my way when it comes to success and happiness? Am I missing out on life itself?*

Don't wait until it is too late to make the changes that matter. Start your kindness journey today. And remember, as a form of love, kindness requires us to pass the torch from one generation to the next. When you're young, kindness gives you the wisdom of an old soul. And when you get a little older, kindness can keep you young.

—Stephen G. Post, PhD
Founder of the Center for Medical Humanities,
Compassionate Care, and Bioethics
Stony Brook University School of Medicine
President, Institute for Research on Unlimited Love
Author of *Why Good Things Happen to Good People: How to Live a Longer, Healthier, Happier Life by the Simple Act of Giving*

Introduction

"Mom, I have something to tell you. But I want you to hear me out before you say anything." When a seventeen-year-old daughter starts a conversation like this, what parent doesn't brace for the worst? Car crash. Date rape. Pregnancy. I held my breath.

Sophie explained that the night before she had been physically attacked by another girl. My daughter showed me the deep nail marks across her chest and back. I flinched. Her boyfriend had taken a photo of Sophie's bloodied nose, which I refused to look at because I didn't want the image imprinted in my brain.

"Are you okay now?" I pleaded.

"Yes."

"Are you afraid something else might happen?"

"Not really."

"Why did this girl jump you?"

"An old grudge… Jealously… A power trip."

My impulse was to bang down the door of the girl's house and ream out her mother. But Sophie took the high road: she showed compassion for the girl and her life situation—much more grace than I could muster. As a clinical psychologist working primarily with girls and women for two decades, I know a lot about mean girls. Sophie didn't expect an apology, and the girl never offered one. It all fell into the past. But the skirmish really stuck in my craw. It was so unkind.

The incident inspired me to dig deep, to draw on my own practices of mindfulness and compassion in the face of meanness. But given the state of our world during the contentious 2016 U.S. presidential election, this felt hard to do. Political mudslinging had intensified. Children had started becoming anxious or aggressive, openly taunting immigrants and disabled people. Bullying was on the rise and so were fears of deportation. Swastikas were painted in our local middle school, and our Holocaust memorial was desecrated. Many of us felt horrified by the discourse on the treatment of women and girls. Moreover, a steady stream of atrocities amplified fears that the world is dangerous: the refugee crisis, terrorist attacks seemingly everywhere, public mass shootings, young black men continually killed by police officers, rioting, acts of retaliation. Instead of responding with unity or cohesion, all this created more divisiveness and separation among us. And yet, I resonated with Selena Gomez's song, "Kill 'Em with Kindness," that my girls were listening to. It was timely, and it kept me going.

The topic of kindness became central to the many conversations I was having at home, among colleagues, and with clients. The question kept popping up, over and over: *What happened to kindness?* The world felt like a meaner place, even though historians insist that we live with less violence and more democracy. Everything seemed to be breaking into pieces, and empathy, respect, and common decency became dwindling resources. Because it can, quite simply, feel too hard to be generous in spirit within such an environment. But as I learned, it does feel easier when you understand the power you are wielding when you are kind. When facing anything that feels threatening—be it political, personal, or as natural as a life change—you can "kill" your fears with kindness. And so the idea for this book was born.

I started inviting people to share reflections and stories on kindness. I'd ask, "Do you consider yourself a kind person? How so?" Either people had a quick reaction: "Oh, trust me, I'm not a kind person." Or they were stumped: "Gosh, I really have to think about that one." Mind you, plenty of people came up with kindness stories: a person giving up an airplane seat for a pregnant woman after her flight was canceled; a cabbie lending money to someone with a lost wallet; an unexpected thank-you card;

daffodils left on doorsteps. But half the time, people had to stop and think for a while about kindness. It became clear that, in general, we are grappling with it—as if kindness is hiding in a corner and we have to search it out to find it.

A Fear of Kindness

We seem to have a kindness phobia on our hands. People are afraid to be kind, to admit to being kind, to trust people who are kind, to pay attention when kindness does happen. What's going on? Has our speedy technological world made it too easy to disengage from one another, fetishize our differences, and normalize a cool-to-be-cruel mentality? Does the world feel so unsafe that we are being overprotective? If so, are we reverting to a survival mentality that reserves kindness for our children, close family, and friends? Have we prioritized values like personal success and happiness over the well-being of others to the extent that the Golden Rule is a relic of the past? Are we expecting people to be kind but don't enact it ourselves? Or do we simply take everyday kindness for granted?

Answer: all of the above. These questions are recycled in history, and different ages have different results. "Kindness is always hazardous because it is based on a susceptibility to others, a capacity to identify with their pleasures and sufferings," write psychoanalyst Adam Phillips and historian Barbara Taylor. Their brief historical account on the rise and fall of kindness laments its decline as the Victorians conceived of it, which was an openheartedness that linked people to one another. There was a sense that our lives depended on being kind because, as philosopher Jean-Jacques Rousseau wrote, "Our sweetest existence is relative and collective, and our true self is not entirely within us." Kindness included expanding beyond people within an inner circle to view strangers as kin, which cultivated the Victorian recognition of a shared humanity. This was uplifting and motivational. It led to caring cultures.

The attitudes of modern society—with emphasis on self-sufficiency, self-interest, and separation—have eroded our confidence in kindness as an orientation to living a meaningful life and as a civic virtue. Kindness has become associated with being weak, fragile, feminine, nostalgic, and untrustworthy. Yet our instinct for kindness percolates to the surface all the time, because our basic neurology is wired to care. As Taylor and Phillips write, it's a paradox: "People are leading secretly kind lives all the time but without a language in which to express this, or cultural support for it."

It's time to create cultures of kindness, at the very least because we are desperate for them. To do so, it's on each of us to transform kindness from a sentimental notion to a natural expression of love, respect, and appreciation for one another. This takes effort. *The Kindness Cure* will give you a language for kindness, encourage you to feel its necessity, help you to make kindness explicit, and give you ways to cultivate it from the inside out. As you will see, you possess a compassionate instinct. It's part of your genetic blueprint. But the capacity for kindness can erode if you don't exercise it, which may be the greatest thing to fear.

Bothering to Care

This fear is made vivid by the Making Caring Common project at Harvard's Graduate School of Education. The executive director, Rick Weissbourd, observes that we are in an era of high-achievement pressure and that we emphasize individual happiness. "What we don't see is more focus on concern for others, concern for community." Weissbourd believes, and I agree, that we need more caring opportunities and demonstrations of kindness, especially among people different from us.

The project's report, titled *The Children We Need to Raise*, surveyed more than ten thousand diverse youth. Almost 80 percent of the youth picked "high achievement" or "happiness" as their top value, while roughly 20 percent selected "caring for others." The general findings were that "young people neither prioritize caring for others nor see the key people around them as prioritizing it." The researchers observed a gap between caring rhetoric and reality, between what parents and other adults say are their top priorities and values and the messages they convey by daily behavior:

> When children do not prioritize caring and fairness in relation to their self-concerns—and when they view their peers as even less likely to prioritize these values—there is a lower bar for many forms of harmful behavior, including cruelty, disrespect, dishonesty, and cheating.

The report cautions us about the misguided values and behaviors of today's children, and it implicates the society in which they are raised. It implicates us. We can see the gap between rhetoric and reality almost everywhere we look: politics, business, education, parenting. If we value caring and kindness, then we need to cultivate it in ourselves, in our relationships, and in our community. Weissbourd reflects: "I think the good news is that lots of kids do value caring and kindness. Just too often they are valuing it second [to happiness and achievement]."

A Rallying Cry for Kindness

Kindness is a neglected virtue in today's culture—but it persists as an instinct we can cultivate as something that benefits us and the world. *The Kindness Cure* is a rallying cry that illuminates kindness as a path to well-being and joy.

Most of us value kindness and have the desire to express it. Doing so inherently takes courage, since we must be willing to experience the vulnerability of connecting with other people—especially those who are different from us. We do have what it takes: it's entirely natural to be relational. Discoveries from a broad range of disciplines—including neurobiology, evolutionary sciences, psychology, and education—offer hope for a kinder and gentler world because, quite simply, we are wired for it and we benefit from it. Studies repeatedly show positive effects of kindness on personal and collective well-being in areas as diverse as physical health, emotional health, relationships, life satisfaction, communities, and even economies. The science confirms ancient wisdom, and the data is piling in. Here are some of the things we're learning that kindness can do:

- Activate emotional regulation and compassion networks in the brain
- Alleviate symptoms related to depression and post-traumatic stress disorder in veterans
- Protect against compassion fatigue in helping professionals and first responders
- Lessen migraines and symptoms of chronic pain
- Promote positive attitudes and compassion toward oneself and others
- Lessen judgment and increase empathy for stigmatized social groups
- Improve body image
- Strengthen romantic relationships
- Improve symptoms related to depression, anxiety, and social isolation in teenagers
- Foster stress resilience and prosocial behaviors in young children
- Promote longevity in those who volunteer

Kindness, it turns out, is a happiness fix. And you have the wiring and the natural motivation to grow kindness from the inside out. With this book, you can strengthen your instinct for compassion and encourage it in others, because when you intentionally practice it, kindness radiates from you. It spreads. It is contagious.

A Guide to *The Kindness Cure*

My deep conversations about kindness reveal that each person wakes up to it in a unique way, often through personal hardship, a shift in perception, an emerging self-compassion, or a desire to create a culture of caring. The stories in this book share these wake-up moments, show how kindness is naturally cultivated, address what gets in the way, and demonstrate how we heal through it. They come from clients, friends, colleagues, social media contacts, and my own experiences; in a few instances, I have changed names or identifying details. Mostly, they are very ordinary stories of everyday people doing the best that they can.

Because cultivating and prioritizing kindness is a daily way of being, this book casts a wide net. After you have read chapter 1, you can jump in anywhere. Each chapter begins with a story, describes a principle, shares wisdom inspired by the sciences, offers a practical exercise to carry forward, and concludes with a reflection. Here is an overview of the four parts that mark distinct phases on your journey.

Part 1: "To Be Human Is to Be Kind" highlights your natural tendency for kindness, caring, and concern as fundamental to survival as a species. This section of the book shows how kindness arises in our everyday experiences. I point out the traps that are so easy to fall into: misguided attention, empathic distress, stress, and indifference toward others. And I show how you can be a kindness warrior.

Part 2: "Your Caring Mind" draws attention to just how wired for kindness you are with your body's amazing physiological map. You can appreciate the ways that this "caring blueprint" operates and understand the tensions that arise in how humans evolved to both survive and thrive. I offer ways to calm your nervous system and regulate your emotions, including empathic ones, because without the burden of emotional upset, you are able to make choices to be loving, kind, and helpful.

Part 3: "Kindfulness" shows you how to strengthen your kind neural pathways through the building blocks for kindness: mindfulness, intention, and self-compassion. Life is full of joys and pains, and responding to challenges with inner strength affects your physical, emotional, relational, and spiritual life. The tenderness and loving-awareness that result have the power to expand appreciation and gratitude for yourself, others, and the world you live in. This creates a life oriented to kindness.

Part 4: "Kindsight" conveys how basic needs for love and belonging are shared by all. How you learn from your past and envision your future depend on how you choose to live in the present moment. With kindness as a life orientation, expressions of compassion, forgiveness, and generosity become contagious—igniting an upward spiral of positivity and well-being for the greater good.

Kindness is both simple and sacred. I hope you, your family, and your friends can be inspired by these stories, share these practices, and thereby cast a wide net for a more loving and compassionate world.

CHAPTER 1

Kindness Takes Effort

A month after my daughter, Sophie, got punched in the nose, I had the opportunity to speak with "the happiest man in the world," Matthieu Ricard, a cherished Buddhist monk, humanitarian, and meditation researcher. I asked him mundane questions about empathy, especially how to help people who feel *too much* empathy. It was a bit of a family curse, I admitted.

"Be glad," he said.

I lamented that it can be exhausting.

"Yes," Ricard replied, "but you start with just one thing, close to home, and that is okay. Better to start with too much empathy than too little," he assured me.

That same week, I visited the U.S. Holocaust Memorial Museum—effectively expanding my exploration of empathy from mundane to vastly cultural. My mother emigrated to the United States from Germany. She was born in 1938, a poor Catholic girl whose earliest memories were of bomb shelters and near starvation. As a child, my mother had no understanding of the world beyond her neighborhood, where she tended goats and helped make ends meet. The revelations of the Holocaust came much later. Once she made her way to America at nineteen years old, she was bearing a heavy burden of cultural shame. My mother overcompensated by instilling in her children an intense sense of obligation, of the self-sacrificing variety, to help others. Like me, my daughters inherited this emotional legacy.

At the memorial museum, I became fixated on a display containing children's books, toys, and posters—all intended to indoctrinate children into militarism, racism, and anti-Semitism. Now anything having to do with the Jewish Holocaust brings me to my knees. And I'm not naive to the repetitions in recent history. But the children's books triggered me that day. I reflected on the vast capacity of the human imagination and began wondering whether—if we can mastermind the spread of cruelty—we can mastermind the spread of kindness instead.

My questions about kindness began to shift, from how hard it feels to how we can grow and sustain kindness. With all that is being hailed these days about *neuroplasticity*—the ability of our minds to be shaped by experience, and the ability to shape our minds with experience—it occurred to me that we can lay down neural networks for kindness. We can engage systematic practices that cultivate feelings of compassion and communities of caring.

Matthieu Ricard describes kindness as a mode of altruism, a form of caring and warmhearted consideration that is manifested in how you behave toward others. For Ricard, your natural empathic concern for others can be a catalyst for kindness and compassion, and this can be expanded through various meditative practices and skills. Empathy is a basic connecting point with others that can be experienced on a continuum: it can cause you to be flooded with distress, or you can harness it into kindness. The empathy I refer to in this book is *motivational empathy*—the ability to put yourself in another's shoes and also not get too lost in heightened emotional states that can lead to disempowerment. I'll explain this more in the next chapter; the key point here is that you can use empathy to intentionally expand your awareness of and capacity for kindness.

What Is Kindness?

Kindness is love in action. So let's first look at love. Your love can be the result of a lasting bond, like you have with friends and family. It can also be expressed for any living or nonliving thing, springing forth at any moment in time. Love is the mother of all uplifting emotions and embraces all other positive emotions—amusement, compassion, gratitude, joy, pride, serenity, and so on—under her care that enrich your life and that of others for the better. Kindness is the *conduit* for the vastness of love's expression. It is any act of love that reflects genuine caring. Such kindheartedness is the embodiment of your feelings of warmth and generosity toward others and the world at large—and your desire to bring relief to those who are suffering. In this way, kindness is both a quality of loving presence and an orientation to life that is intentional and active.

This is a message offered by spiritual masters, mystics, and poets: love and kindness are within you, they connect all of us, and they're also larger than us. When such limitless *loving-kindness* is practiced, your relationships with yourself and others lead to authentic happiness and well-being. It is an antidote to fear. Sharon Salzberg teaches how this works: a mind filled with fear can be penetrated and conquered by loving-kindness, and a mind that is infused with loving-kindness cannot be overcome by fear. This kind of unconditional love toward yourself and others is called *mettā* in Buddhist texts. "Mettā—the sense of love that is not bound to desire, that does not have to pretend that things are other than they are—overcomes the illusion of separateness, of not being part of a whole." This type of loving is an expression of kind acceptance. Rather than pushing away difficult feelings or reacting aggressively to change something or someone or the world itself, you can experience a profound happiness that does not depend on external conditions. But the world is constantly testing us, isn't it?

Under a SPEL: How Stress Erodes Kindness

When you are disillusioned, afraid, or feeling threatened or unsafe, it's hard to engage your kindness instinct. You can become exhausted, indifferent, and uncaring—all states that cause you distress. The natural response to stress is to go into a protective mode. The brain's internal alarm system rings so loudly that your only options are *Fight! Flee! Faint! Freeze!* If it's a matter of survival this is important, as we are built to react quickly to stressors. But when your alarm system switch is stuck so that it's always on, your body can't recover—and stress becomes chronic. You feel worn down physically, mentally, and emotionally in a state we'll discuss more in chapter 10.

Experts explain that the stress response and its attending negative emotions *narrow* your focus to immediate action. You can become small-minded and mean without even realizing it, as you snap at people, become overly or undeservedly critical, and in general share your negative outlook. On the other hand, positive psychologist Barbara Fredrickson tells us that positive emotions *broaden and build* your inner resources over time, so they serve you well, especially in hard times. They trigger the body's self-soothing system, and you feel safer and calmer. You become able to rest, restore, and revise your mindset toward expansiveness, generosity, love, and kindness.

In a chronic state of personal distress with a narrow view, it's hard to arouse the empathic concern that will help you get perspective on things and be kind toward yourself and others. This erosion of empathy isn't intentional by any means. Stress is like a slow-dripping poison that gradually weakens and disempowers your capacity for empathy and, therefore, for kindness.

I call this effect ***S**elf-**P**rotective **E**mpathy **L**ethargy*, or SPEL. It creeps up when we are consumed with ourselves: too busy, tired, afraid, overworked, overwhelmed, tethered to distractions like technology, or simply burned out: basically, too consumed with coping to care. Any of us can get caught under a SPEL, for internal or external reasons. It all comes down to how you respond to the things that trigger stress. Because when you respond to stress well, you awaken from the SPEL, and you realize you can handle any challenge. *The Kindness Cure* guides you toward this awakening.

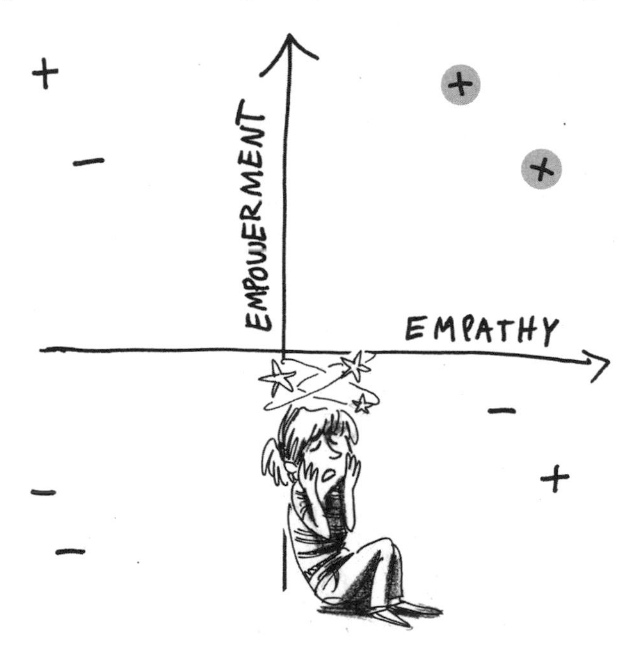

You get caught under a SPEL when there is an imbalance between empathy and the extent of control, or empowerment, you feel to do something about it. When stressed, overwhelmed, or lacking resiliency, you have fewer inner resources—and therefore less motivation—to feel or be kind. The antidote is to move into a sweet spot, as conveyed on the grid above. Being aware of where you fall on the Empathy-Empowerment Grid can help you reflect more deeply about what might help or hinder kindness in you.

+/+ **High empathy/high empowerment** is a state of caring that results in kind behaviors and a personal commitment to the well-being of others. It is a feeling of empathic concern that is followed by acting in alignment with that feeling to offer things such as a caring presence, affection, or relief. This generosity becomes contagious, the circles of care widen, and social justice is valued. This orientation brings a healthy striving for both the personal and collective good.

+/– **High empathy/low empowerment** is being under a SPEL. The caring impulse toward yourself or others is present, but the ability to be of help and service, or to affect positive change, is short-circuited by real obstacles or personal discomfort, fear, exhaustion, or disillusionment. You are more likely to be in a state of empathic distress or compassion fatigue, which is the inability to tolerate the pain or suffering of others.

–/+ **Low empathy/high empowerment** is a position of power and confidence. You are motivated by selfish goals, act at the expense of others, harbor a disregard for the well-being of others, and view others as outsiders.

–/– **Low empathy/low empowerment** is a state of apathy and isolation. This may happen when faced with a real or perceived threat to survival, pervasive trauma, disengagement from others, or an ingrained sense of learned helplessness. You feel numb, so you can't act.

Breaking the SPEL by Opening Your Heart

Being kind requires the willingness to open your heart, and it also requires a conscious and sincere effort to be kinder than you already are. Even though you possess a compassionate instinct, your capacity for love and belonging can erode if you don't exercise kindness. To offset erosion, work your way back to kindness. Warding off a SPEL takes effort, courage, and skill. Fairy tales and cultural narratives tell us that spells can be broken—by a true love's kiss, enchanted cloaks, ruby slippers, or enacting feats that are brave, honest, kind, and selfless. Ultimately, these stories are about waking up to the human experience of love.

There's an elixir of six ingredients that can be blended to break a SPEL and ignite or rekindle your kindness. You can remember this recipe with the acronym PEPPIE—like a vital energy force. Sprinkled throughout this book are PEPPIE examples and exercises that'll inspire and help you to build a potent reserve of these skills. Reflection questions accompany each description to give you a taste of what a kindness cure entails.

Presence. You can cultivate this by learning to feel grounded in physical space and in your body, and by tuning in to moment-by-moment experiences that include sensations, emotions, and thoughts. This is a fundamental practice of centering yourself with kind awareness, or *kindfulness*, even when life feels hurried or overwhelming. It's like discovering a golden egg—a vital source of well-being.

> *What happens when I follow my breath in and out? What feelings and bodily sensations do I notice right now? What thoughts are streaming through my mind? What do I need to feel safe and supported? Can I notice the moments in my day, the pleasant and unpleasant, without judgment?*

Emotional regulation. As you engage your body's natural self-soothing system when feeling upset or aroused, you strengthen your ability to recognize and manage a wide range of emotions—including empathy, a precursor to kindness. The ability to manage emotions fosters social-emotional intelligence, enriches relationships, and results in overall well-being.

> *Can I find my center even if my emotions are triggered and my thoughts are scattered or scary? How can I walk in another person's shoes and*

not be overcome by emotions? Can I find comfort and support when I need them?

Keeping **P**erspective. This entails becoming curious about life experiences, learning from them, and imagining new possibilities for kindness, connection, and growth. To see the world in this way is to cultivate your inner coach, wise mind, or higher self. This allows you to expand your field of awareness because you see through the lens of understanding and kindsight.

Can I listen with compassion? Can I contemplate others' points of view, finding common ground even when I disagree? What is the best way I can be of help when I see suffering everywhere I look?

Purpose. When you feel purposeful, you believe in yourself and others, imagine possibilities, find new solutions that bring benefit, identify your values and guiding principles, and act on those principles. A purposeful life gives you the security in yourself to connect with others in open, fearless ways. This emboldens your integrity and strengthens empathy, which allows you to experience and support beauty and dignity among humankind.

What is important to me? What are my values? What possibilities do I imagine? What does a meaningful life look like? Could I make it part of my purpose to be a kindness warrior?

Integration. Integration nurtures healthy habits for mind, body, and spirit. A wellspring of inner stable and benevolent strengths will grow as you integrate habits such as learning to appreciate the good moments, feeling safe, caring for yourself, practicing forgiveness and gratitude, listening with compassion, and setting healthy boundaries. Such skills build resilience to bounce back from inevitable setbacks or unforeseen obstacles, so you can perceive the wholeness of life's joys and sorrows.

What are the tried-and-true ways in which I nurture myself with love and kindness? Are there other things I can do differently to demonstrate how I care for myself and other people?

Effort. Remember, kindness is love in action. Harness your empowered senses of personal agency, creativity, and inner strength for the good. These attitudes naturally want to engage in efforts that create kindness, compassion, and social justice. You can consciously exert your personal power and take leaps of faith, even when outcomes are uncertain and failure is possible.

What can I put joyful effort into? How can I be of service to others? Is there a small or big kindness that I can do for myself or someone else? What might I like to try out, create, or collaborate on? How can I connect with others not like me? How can I put love into action?

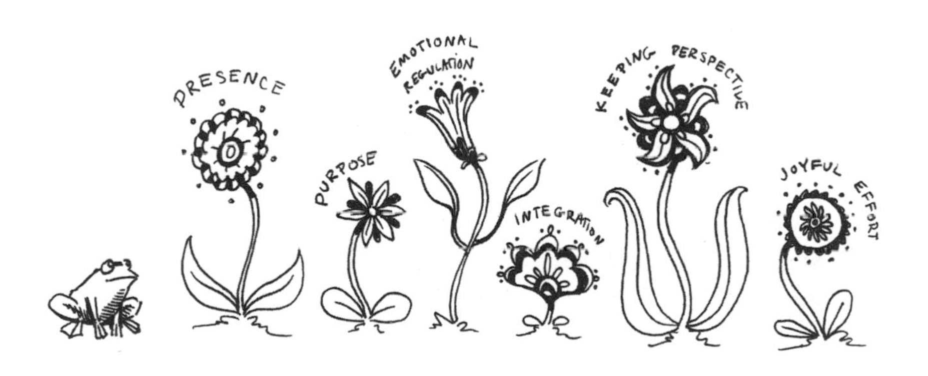

As you can see, some of these ingredients are inner ones and some are actualizing ones. The chapters that follow offer you the skills to grow kindness from the inside out.

PART 1

To Be Human Is to Be Kind

CHAPTER 2

Your Kindness Instinct

The world can at times feel like a very unkind place. The year my daughter, Josie, was twelve years old, three horrendous events occurred: the "Batman Shooting" at a movie theater where twelve people were killed and seventy wounded; the massacre at Sandy Hook Elementary School that left twenty-six children, teachers, and staff dead; and the Boston Marathon bombing that wounded hundreds of people and killed three. It was a year of extreme tragedy. Despite my efforts to minimize overexposure, Josie felt a real and strong fear that these things could happen to her. Lockdown drills were being done regularly at school, and she began imagining herself and her loved ones torn to bits. Josie's heartbreaking response was to wonder aloud, "*Why* do people do this?" A basic dread took hold of her.

Then something sparked an awakening. In a long-jump sandpit, some kids discovered a pair of baby field mice. Josie became determined to rescue them, even though everyone knew it was a lost cause and coldly said as much. She declared, "I know they will die. At least I can give them comfort until they do."

At her insistence, I drove Josie to the sports field as she clutched a bottle of Pedialyte and a dropper. She dashed across the track to be Nurse Nightingale. The mice were still alive, and she gently placed the pink creatures in a box of tissue. As I watched, it hit me: Josie was fully alive with her compassion for all living things. It was the act of caring itself that mattered to her. As if, by demonstrating the kindest way that she could possibly act, she could turn the world around.

She gave them names as children do—Bradley and Charlotte—and carefully nursed them. One mouse died a few days later. Josie held out hope for the other but, in the end, she buried the pair in the garden. While she was frustrated that they hadn't lived, she consoled herself with this: "At least I tried." We were all moved by her efforts. At a time when the world felt burdened by grave issues, saddled with grief, frustration, and helplessness, Josie found healing and empowerment by demonstrating her own kindness. She showed a brutal world that she would care for the meekest of its creatures. And, in doing so, she discovered that what feels best is not evading harsh realities—it's being kind along the way.

Being a Kindness Warrior

True kindness does not have an agenda or ulterior motive; it is, as Josie showed, an instinctual response that can feel highly energized and even fierce. Yes, kindness can be fierce. Chögyam Trungpa, a Tibetan Buddhist teacher, goes so far as to say that a compassionate mind is a warrior's mind. "To understand our self-nature, as well as the self-nature of humanity, we should focus on what has beauty and dignity among humankind. Doing so will rid you of a fearful mind, and change it into a spiritual warrior's mind." To overcome your fears of living in a world where painful things happen, expand your compassionate nature. Because it is innate. Know that what you do matters to others, so be *caring* and *careful* about your actions. Be a kindness warrior. Encourage the pure energy behind this instinct.

Honestly, our family saw Josie's efforts as completely pointless. We wanted to protect her from inevitable disappointment. We were buzzkill, as she said, and we were indeed killing kindness. The truth is that joy and suffering go hand in hand, and it's a disservice to shield children and each other from either.

Super-Caregiving Species

Our biological instinct for kindness comes together with our social conditioning to inform how we engage in the world. You have a deep instinct to care. We all do. Humans have evolved not just to survive but to thrive. There is a lesser-known aspect of the theory of human evolution: sympathy. Stronger than self-interest or self-protection, sympathy is a reflexive social instinct. It developed from our need to care for vulnerable babies, who require years of nurturing. This made us a super-caregiving species, which—from one generation to the next—rewired and refined our nervous systems. Compassion and kindness are so much a part of the human blueprint that they are "embedded into the folds of our brains," as the researcher and academic Dacher Keltner puts it. Scientists and spiritual leaders agree that these distinctly human qualities are built into our cellular blueprints and woven into our spirits. We are wired to care.

How this natural propensity for kindness flourishes is largely up to you. The psychologist and neuroscientist Richard Davidson said, "Human beings come into the world with innate, basic goodness. When we engage in practices that are designed to cultivate kindness and compassion, we're not actually creating something *de novo*—we're not actually creating something that didn't already exist. What we're doing is recognizing, strengthening, and nurturing a quality that was there from the outset." And therein lies the hope and the responsibility.

How you express this exceptional capacity depends on your unique life experiences, because the reality is that aggression and competition are part of our evolutionary legacy too. So is fear. Fortunately, your hardwiring can be trained through efforts to nurture and strengthen your kinder instincts. You can grow in kindness. It begins with an awareness of your kind nature, and it culminates in the determination to honor that nature in the *Purpose* you elaborate for yourself and the *Effort* you engage in.

Kindness in Practice: *Rekindling Kindness*

I invite you to become a kindness warrior. The process begins by simply thinking about what kindness means to you. Consider your own thoughts, feelings, images, and aspirations. Take out a journal and complete this reflection exercise.

> Call to mind at least three instances when you experienced kindness. These could be moments when your kindness instinct caused you to override hesitation and show care or concern for the well-being of others. They could include times when you were on the receiving end of a kindness or when you witnessed someone else putting love into action. You could even include stories of kindness that you heard about that have stuck with you. Here are some prompts to help get you started:
>
> I remember when I helped…
>
> I was reminded about human kindness when…
>
> I'll never forget when __________ was kind to me…
>
> When I think about kind people in the world, the list includes…
>
> When I think about kind people in my life, I call to mind…
>
> A time I stood up for kindness was…
>
> When I think about when I had to get out of my comfort zone to be kind, I remember the time…

By noticing your instinct toward kindness and compassion, you can begin to kindle it—or *rekindle* it. It can be like a campfire: you start with the smallest of sticks and blow gently on the flames. Then the logs catch in an outpouring of warmth and light. So recognize kindness when you see it. Notice that, like Josie, many of us are warriors without even knowing it.

Reflection: *Kindness is my natural state. I ask myself: What has life taught me about kindness? In what ways am I a kindness warrior?*

CHAPTER 3

Happy to Help

Jonah and Dylan have been friends since preschool. When they were in first grade, Dylan's mom told Dylan that Jonah has a rare liver disease with no known cure. Dylan's response was to help. He decided to raise money for the research needed to find a cure. His idea was to draw a colorful picture book and sell it at a school event. The book was called *Chocolate Bar*, which was code for "awesome." It listed "chocolate bar" things such as going swimming and visiting Disney World. At Dylan's insistence, his parents made two hundred copies, and Whole Foods Market donated one hundred chocolate bars to the cause.

Both the books and bars sold out, and Dylan raised $5,000 in a few hours. Inspired, this six-year-old set a goal: raising one million dollars. His mother was shocked. Did he even know what one million dollars was? No matter. Dylan kept at it. Friends told friends. Their story was picked up by local and national news.

The boys were a sight to see. Dylan has dark hair and eyes, deep dimples, and an impish grin. Jonah is fair, red-headed, with his front teeth still growing in at the time. Together, they could be pals on *The Little Rascals*. But their mischief turned miraculous: in just two years' time, with tireless assistance from their parents, the book raised a million dollars. Because of their efforts, Jonah's doctor started clinical trial research that may someday lead to treatment or cures for many liver diseases. Dylan's optimal zone of high empathy and high empowerment will benefit countless people in the future.

Empathy Expands Your World

The word *empathy* means "feeling into" something or someone. Kids these days like to say, "I feel ya," whereas adults are more likely to say, "I hear you." Empathy engages your imagination as you sense the thoughts, feelings, and intentions of other people. It thrives with nurturing experiences and secure attachments to caring people. It is also shaped by trial and error, discernment, and by knowing personally and deeply what it feels like to be hurt and to do the hurting.

Empathy is a necessary starting point. We are wired for empathy and endowed with the mental capacities to harness it. The roots of empathic resonance can be seen in the emotional contagion of babies, when one laughs or cries in response to another. We experience it with a human cry, voice, movement, or facial expression. As we grow, it becomes possible to vicariously share in another's emotional or mental states without being lost in them and mistaking them for our own. Psychologists have assigned these attributes and skills to two basic functions in the mind:

- *Emotional empathy* is when you instantaneously feel happy or sad when another person feels happy or sad. You mirror the emotional states of another person, tapping your neural circuitry for emotional response.
- *Cognitive empathy* gives you the ability to take on the perspective of another person, imagining or intellectually understanding their thoughts and feelings without necessarily evoking emotions. This is referred to as *theory of mind* or *mentalizing*, and it engages a different neural circuitry in your brain.

We are all wired and raised differently, with unique sensitivities, personalities, and attachments that affect the development of empathy. Both emotional and cognitive aspects of empathy lead to *motivational empathy*—empathy that motivates us to care about each other and to act kindly. We need these three aspects of empathy so that self-awareness, emotional regulation, and gaining perspective—all core ingredients in PEPPIE—can develop. While helping is child's play, as Dylan showed, we need this maturity so that empathy leads to intentional helping, collaborative, and caring behaviors.

Growing Up Empathic

Empathy grows as we grow. In a short time, as children begin to understand their own emotions and to know that other people also have their own emotions, empathy leads to caring actions. Toddlers as young as fourteen to eighteen months spontaneously show kind, helping behaviors, such as bringing a bottle or toy to comfort another toddler in distress. Children readily enjoy helping others and, without thinking about it, derive pleasure from doing so.

The work of Felix Warneken suggests that the roots of altruism can be seen in toddlers' spontaneous helping behaviors. He sets up scenarios with a bumbling adult pretending to have trouble with a task, such as putting away books in a cabinet or reaching for a pen on a desk that has rolled out of grasp. Toddlers infer the goals and intentions of the grown-up, and then they open the cabinet door or return the pen. Moreover, children don't necessarily look for rewards. This speaks to an intriguing finding in many studies that rewards can actually thwart empathy and undermine nice behavior. This seems entirely counterintuitive, yet it's true: if children get rewards—stickers or tokens, for instance—for helping, they are less likely to help in the future. Why? The natural delight a child experiences by helping someone (the intrinsic reward) is undermined by the prize (the external reward). Better to express heartfelt joy and appreciation, because children are simply happy to help.

Kindness in Practice: *An Emotion Vocabulary*

You may be wired for empathy, but it needs to be encouraged and developed. To cultivate emotional empathy *and* cognitive empathy, begin with *Presence*—with an awareness of how you feel. Soon you'll notice that you're better tuned in to how others are feeling too.

Studies show that labeling emotions decreases reactivity and allows for recognition and learning to happen. Labeling emotions is something you can do so that you are neither overcome by emotions nor numb to them, whether positive, negative, self-focused, or other-focused. When you practice turning feelings into words, you develop a vocabulary for them, which expands your awareness of emotions—both your own and others'.

> During the next seven or so days, before bedtime, reflect on your day and identify a situation that evoked a feeling—pleasant or unpleasant. Write your feelings down in your journal. Stretch your vocabulary to capture exactly how you felt using the following word list. If there's a subtlety that isn't captured in this list, go online for feeling inventories, use sentences or word combinations, or invent a word. Be curious. Note any surprises or patterns that arise during the week.

Active
Affectionate
Afraid
Agreeable
Amused
Angry
Annoyed
Ashamed
Awesome
Awful
Blue
Bored
Brave
Calm
Capable
Caring
Cheerful
Clumsy
Comfortable
Confused
Cooperative
Creative
Cruel
Curious
Depressed

Determined
Disappointed
Discouraged
Disgusted
Eager
Ecstatic
Embarrassed
Enthusiastic
Excited
Fantastic
Fearful
Fed up
Focused
Forgiving
Freaked out
Free
Friendly
Frustrated
Generous
Gentle
Gloomy
Grateful
Guilty
Happy
Hopeful

Ignored
Impatient
Important
Inspired
Interested
Irritable
Jealous
Joyful
Jumpy
Lively
Lonely
Lost
Loved
Loving
Mad
Miserable
Moody
Nervous
Optimistic
Overwhelmed
Passionate
Peaceful
Pleasant
Proud
Rejected

Relaxed
Relieved
Sad
Safe
Satisfied
Scared
Sensitive
Serious
Shy
Sleepy
Stressed
Strong
Stubborn
Tense
Thoughtful
Thrilled
Tired
Troubled
Unafraid
Uncomfortable
Weary
Worried
Yucky
Zany

Empathy is a bridge between the emotional and the cognitive—our heads and our hearts—and it makes acts of both reason and love possible. Without this bridge, kindness is blocked, and you fall under a SPEL. Our emotions are a primal connection with one another. Recognizing your emotions—pleasant and unpleasant—opens pathways for recognizing them in others. This makes ordinary and extraordinary acts of kindness possible.

Look what happened when Dylan learned about his best friend's illness. His empathy for Jonah led to actions that gave hope to his friend and to many others who suffer similarly. The natural examples children offer can help us see our own possibilities.

Reflection: *Feeling and understanding others' experiences begins with feeling and understanding my own.*

CHAPTER 4

Cultivating Courage

"I've slept on these steps before. Me and the library, just taking a nap together." A homeless guy, Michael, was giving a tour of our city—through his eyes.

"This made me look at the streets differently," reflected Sophie, my oldest daughter, "because they really shouldn't be someone's home."

Her friend Sabrina added, "The first moment we met Michael was so jarring. I was very shy and hesitant to speak to him, but now I see he's such a normal, wonderful person."

When Sophie and Sabrina were in eighth grade, they participated in an overnight urban outreach program called City Reach. They gathered to share hospitality, service, and reflection. Homeless people gave walking tours of city streets and answered any questions the kids had, an encounter that was both awkward and intimate. Five years later, I asked the girls what had really stuck with them.

"People told us their stories and it changed the stigma," remembered Sophie. "They're not all drug addicts. Some had normal lives before one misfortune struck after another. Most were nice and funny and personable. Some were facing really hard circumstances."

"When they explained how they find food and places to sleep, I was hit with the reality of their daily hardships," recalled Sabrina. "Before, homelessness was a faraway problem that didn't affect my life. But by literally walking the paths they live daily, I felt how real the problem is."

The girls were still moved by the experience. "It's painful to imagine myself or any of my loved ones in that scenario," Sabrina reflected. "It's a heart-wrenching experience that I would never wish upon anyone."

"Michael told us something that really stuck with me," Sophie added. "He said that, when you are homeless, no one gets you. So everyone avoids eye contact. He said that if one person would just smile and say hi, he would feel better about life. My lesson was that even if I don't give money, I can give a smile. Homeless people aren't invisible."

In high school the girls became student leaders, educating peers about homeless veterans and fund-raising to help them. Now in college, Sabrina says, "The experience in eighth grade definitely shaped my identity as a young adult. I realized that so much of the world needs help, and I now plan to contribute to fixing their challenges." As Matthieu Ricard said to me, when empathy starts close to home, it can become compassionate action.

To Walk a Mile in Another's Shoes

Understanding the experience of another person is an adventure of love and imagination as you think and feel your way into their shoes. This takes courage because you will face pangs of judgment and internal conditions that block your kinder nature. You will step out of your comfort zone to witness another's vulnerability—and your own.

When you reach deep within to truly understand another person, vulnerability arises. "Experiencing vulnerability is a choice—the only choice we have is how we're going to respond when we are confronted with uncertainty, risk, and emotional exposure," writes Brené Brown. You can experience fears of association ("I'm not like them") or rejection ("I don't belong") or unworthiness ("I'm not good enough for them"). Uncomfortable feelings can arise: anxiety, disgust, heartache, or embarrassment. Because by reaching out to others, you expose yourself.

Erika Lantz, producer of the *Kind World* radio series, puts it this way:

> Sometimes a little kind act is very small and it really doesn't do anything to disrupt your day. It can be just a split second and you're actually having a positive impact. Other times it does cost something to be kind. Sometimes it takes a lot of time. Sometimes it inconveniences you. It takes some of your emotional energy or just your physical energy. You have to be vulnerable to ask for kindness; you have to be vulnerable to talk about it. You also have to be vulnerable to show kindness.

This is the challenge of being kind.

Rick Weissbourd observes that "A lot of us struggle with courage, like when to be an upstander, when to stand up, when to stand down, who to stand with, what to stand for." To cultivate this courage, he believes we need more capacity for "our ability to manage envy, jealousy, shame, and our appreciation of other people." Once more people have this capacity, he believes we will all benefit, personally and societally—especially our children.

Facing Vulnerability

Today's societal pressures and attitudes reinforce independence, competition, social comparison, self-absorption, and personal achievement. They encourage feelings of separation and fear of other people. As the educator and activist Parker Palmer wrote:

> The instinct to protect ourselves by living divided lives emerges when we are young, as we start to see the gaps between life's bright promise and its shadow realities. But as children, we are able to deal with those "dark abysses" by sailing across them on the "wingèd energy of delight" that is every child's birthright gift.

He is referring to a line in a poem by Rainer Maria Rilke about crossing unimagined bridges. The bright energy that children have comes from *the soul.* Palmer points out that as we grow up and "cross the rising terrain between infancy and adolescence…we lose touch with our souls and disappear into our roles." We begin to live divided, separated from each other, and become "masked and armored adults." Of course, this is not good for our souls or for humanity. We need to cross bridges by leaping into moments of connection and vulnerability, like Sophie and Sabrina did, choosing to open our hearts.

Feeling connected, supportive, and supported means stepping beyond momentary comfort and taking risks to reach out. To be kind means you must cross relational space between yourself and others, which is filled with uncertainty. You will ask: *Do I approach or avoid? Do I close my heart or expose it?* It's easier to put yourself in others' shoes when you have something in common, have had a similar experience, or share a point of view. But what if you don't? Is that reason to continue being separated? Or can you find the bridge made by a common humanity? Along the way, it helps to clarify your own feelings, thoughts, beliefs, and values, as the girls in the story did. It's a process.

Kindness in Practice: *Self-Awareness Breeds Courage*

In the practice of kindness, there is a slippery point at which showing empathic concern and enacting kind deeds could go one way or the other: toward discomfort and distress or toward ease and joy. The direction depends on a sense of safety and your balance of empathic concern with how empowered you feel.

To explore your experience, revisit the Empathy-Empowerment Grid in chapter 1. Pull out your journal and divide a page into three columns, as shown in this sample. At the top, name a painful situation people find themselves in. Then write down the feelings, thoughts, and reactions that the situation brings up.

A painful situation that worries or stresses me is: ***homelessness***

My Feelings	My Thoughts	My Reactions
Fear, sadness, hopelessness	***What if this happens to me? I wonder whether they are all drug addicts.***	***I'm scared to face them. I feel guilty that I have a family and a home.***

Where I fall on the Empathy-Empowerment Grid: ***high empathy/low empowerment***

This exercise is simply about noticing your discomfort, preconceived notions, or judgments. I encourage you to do it whenever you are at the slippery point of kindness to cultivate the self-awareness you need to answer these questions:

In what situations am I caught in a SPEL?

What makes me uncomfortable about other people?

What do I need to feel safe or supported?

What conditions influence how empathic and empowered I feel?

What are the risks of identifying with others, especially with someone who is different from me?

What thoughts and feelings lead me to turn way, feel aversion, or feel numb?

Who am I leaving out of my kindness circle?

How can I bring kindness to this moment?

What are alternative responses?

When you intentionally cultivate love and kindness, respect and understanding, you will begin to dispel fears so that you can be propelled by courage. This gradually creates conditions for ever-more kindness to thrive in your everyday life. Father Greg Boyle expresses this beautifully: "Our common human hospitality longs to find room for those who are left out. It's just who we are if allowed to foster something different, something more greatly resembling what God had in mind. Perhaps, together, we can teach each other how to bear the beams of love, persons becoming persons, right before our eyes. Returned to ourselves." Being joyful, authentic, and wholehearted with each other takes courage, and giving this to ourselves and others is a powerfully kind thing to do.

Reflection: *Whatever I do today, whoever I am with, I imagine that kindness is lighting my way.*

CHAPTER 5

Compassionate on Purpose

Ever been sucker-punched by life? It can take your breath away in an instant. A cancer diagnosis feels like that. When Christine was told she had breast cancer, she was shocked because she thought she had done everything right. She had led a healthy lifestyle, practicing yoga, eating vegetarian, and running marathons. She had even been a health coach. Instantly, she was forced to undergo something she abhorred: conventional medicine's regimen for potentially life-saving but toxic treatments. It was an anonymous, cold, sterile experience.

"Radiation treatment is very odd," Christine explains. "You go in this tiny dressing room to get changed into a robe. Then you go into the waiting room while the previous woman is in the radiation room. One woman comes out and the other goes in, but you never see people. It's a silent, secret society."

While Christine understood that the protocol ensures patient privacy, she felt she couldn't be the only woman experiencing how impersonal it was. "The changing room had a tiny mirror in it," she says. "I started leaving sticky notes on it with inspirational messages. I had the first appointment of the day, so each note was up all day long. When I returned the next day, I would take that quote down and put up a new one. Doing this helped me refocus my brain before going into radiation, and it became a collaborative effort with my kids. They came up with fun, motivational messages. It was really meaningful. One of my favorites was 'Do your best and forget the rest.'"

Christine left one note for nearly each of the thirty-three treatments. "I never got to see who read the messages. It was our little secret, like a covert operation, for me and the kids. And in such a hard time, it brought us some measure of joy."

Love in Action

The Dalai Lama famously advises, "If you want others to be happy, practice compassion. If you want to be happy, practice compassion." Christine and her children may have derived more pleasure coming up with inspirational sticky notes than the women who read them. The benefit to her was enormous: she avoided falling under a SPEL by feeling empathy and acting with empowerment.

What Christine did was exhibit compassion. Whereas empathy applies to a broad range of experiences and can be expressed in times of both joy and sorrow, compassion is specifically directed at another's heartache and pain. The very word, derived from Latin, means "to suffer together." While compassion involves a desire to alleviate the suffering you perceive, this is sometimes possible and sometimes not. At the same time, compassion can be a motivating force behind selfless acts of kindness and generosity. *Altruism* can be defined as "compassion in action," doing things in the service of another's well-being. Mother Teresa said, "If you are kind, people may accuse you of ulterior motives. Be kind anyway." And it can seem like an ulterior motive to make yourself happy by doing good for others. When Christine wrote those notes, she was lightening a painful treatment experience—as were her children, who had to cope with fears of losing their mom. But they were motivated by compassion for others and the desire to help. It's likely those kind notes helped other women warm up the cold experience. You can only ever guess at the effects of your kindnesses, which is selfless love in action.

The Elements of Compassion

Science backs up the Dalai Lama's simple statement about compassion. Through brain imaging, neurologists have shown that when we observe someone giving money to charity, the pleasure centers in our brains are equally active as when we receive money ourselves. And our sense of well-being is way higher when we give to others than when we spend money on ourselves. In other words, "the joy of giving" has an anatomical basis in the brain. Scientists view compassion as an exquisite capacity among humans, one that leads us to perform not only acts that ensure survival but also ones that make us feel good. They suggest that compassion involves the following five elements:

- You recognize that another is suffering.
- You understand that human suffering is universal—something all humans can and do experience, and therefore something to empathize with.
- You feel empathy for the person suffering.
- You tolerate others' uncomfortable feelings (such as anger, distress, fear, and shame) and allow a measure of your own so that you can be present for the person who is suffering.
- You take some action to relieve the suffering.

These elements may help you see that compassion is not something untouchable, precious, or reserved for spiritual masters. Even now you might recall a situation or event that tugged on your heartstrings and awakened one of these elements in you. Consider the previous chapter, in which youngsters confronted homelessness in their city. Compassion is instinctually residing inside you, like them, just waiting to be brought out and honed.

Kindness in Practice: *Loving-Kindness Meditation*

Your compassion can expand, as it is shown to be a skill that you can learn through interactions with others. Another proven way to grow it is through meditative techniques that contemplative traditions offer. One ancient, lovely core practice from Buddhism is loving-kindness meditation. It is so simple that anyone can do it, even young children, to cultivate the wish that others be happy. It is like a blessing or prayer. Psychologists use loving-kindness meditation to develop wellness skills to enhance positive emotions. Studies show beneficial effects in a variety of people, including those suffering from work stress, anxiety, depression, post-traumatic stress disorder (PTSD), and schizophrenia. This is because imagining and directing love and kindness toward others makes you feel better even in the face of difficulties. A loving-kindness meditation cultivates compassion from the inside out. Rather than focusing on pain or suffering, the practice begins with you opening up to feelings of love, affection, and friendliness toward something or someone. Here are instructions you can follow:

1. **Sit quietly.** Begin by sitting with a straight posture, and relax into it so that the posture evokes a sense of grace, strength, and dignity. Gently place your hands on your lap or, if you desire, over your heart. Tune in to your natural breathing for a few minutes.

2. **Focus attention on your heart space and body.** Place your attention around your heart area, in the middle of your chest, perhaps repeating words such as "love," "peace," or "warmth." As you say this, envision someone or something you

feel caring toward. It could be a child, loved one, pet, or comforting object. This ignites feelings of affection and love. Let these feelings radiate through your whole body, holding you in a warm embrace.

3. **Focus on phrases that evoke tenderness for yourself.** Feel the sense of caring, love, and healing wash over you. Softly say to yourself any of the following phrases, and explore how they resonate within you.

   May I be well.

   May I be healthy.

   May I be happy.

   May I feel at peace.

   May I feel safe.

   May I be at ease.

   May I feel loved and cared for.

4. **Expand your tenderness to a loved one.** Envision a loved one or someone who evokes affection or respect. Offer them feelings of warmth and caring, and wish them well with words like these:

   May you be well.

   May you be healthy.

   May you be happy.

   May you feel at peace.

   May you feel safe.

   May you be at ease.

   May you feel loved and cared for.

5. **Wish a stranger well.** Extend your warm feelings to someone you do not know, someone you feel neutral about, like a fellow commuter on the subway, a person crossing the street, the checkout lady at the supermarket. Repeat the words offered in the previous list. *May you be well…*

6. **Include people who challenge you.** Call to mind someone you struggle with a bit. It can even be someone you find irritating, such as an in-law, coworker, or telemarketer. At first, do not focus on someone who has hurt you. With practice, over time, you may become ready to add this person to your expanding circle of warmth. Repeat the words to wish him or her well. *May you be well…*

7. **Extend wishes of well-being to the entire world.** Broaden the warm wishes to include your greater community, whether a church group or a city, and then include the world at large.

8. **Return to your body, yourself, and your life**. When you are ready to return, repeat the phrases that wish yourself well. Close the practice by gently letting the feelings of loving-kindness ease and then paying attention to your breathing. Slowly open your eyes.

Like Christine's notes, some messages of loving-kindness will stick and some won't. Ultimately, what matters most is the intention and affection you put into them. You can ignite positive feelings toward others by wishing them well, which will, over time, lead to acts of kindness and compassion. A loving-kindness meditation can become a daily prayer, a moment of appreciation for another person, or a personal practice to train your compassion muscle. The beauty of this practice is that it cultivates feelings of warmth and tenderness without going anywhere or doing anything, instantaneously evoking your awareness of being part of one big human family. From there, see how you are inspired to act.

Reflection: *I choose to feel loving-kindness. I open my heart wide to all of humanity.*

CHAPTER 6

Imagination's Objects of Affection

When word got out that I was interested in kindness, my social networks turned into kindness spotters. I was tagged on, messaged with, and e-mailed stories constantly. It became obvious that, once you look for kindness, you see it everywhere.

One viral video, "Sewing Hope," hooked me completely. I watched it dozens of times, shared it, and made every kid who entered my home check it out. It is about a twelve-year-old Australian boy named Campbell who makes stuffed animals for sick children. He sews these creatures himself and, in three years, he's given away more than eight hundred of them.

Campbell was originally inspired to buy Christmas toys for sick children, but because he has eight siblings there was no extra money for presents. Campbell's solution? He decided to make them. He borrowed his mother's sewing machine, even though he had no idea how to use it. He Googled patterns for stuffed animals. His first, a "little, ratty, wiggly bear" took him five hours. Many bears later, Campbell can make an animal in one hour, creating fantastical designs worthy of Jim Henson's Creature Shop. His goal is to make one animal per day, 365 days a year.

Campbell's mother says, "He looks at sadness and tries to turn it upside down." He prefers sewing creatures to the usual after-school activities, such as playing video games. He can't get enough of it. "If I say, 'Don't do it,' he sneaks," laughs his mom. In the beginning, Campbell did little jobs to earn money for fabric—until word got out on social media and people like me started making donations to cover his supplies.

Campbell visits a local children's hospital to personally deliver the stuffed animals. He engages the children with friendliness and curiosity, which draws them out. He listens, affirms, and asks simple questions better than most psychologists or doctors with years of training. When he visited one girl being treated for brain tumors, he asked, "How are you feeling?" She answered that she was good and that her bear, Cherry Roseberry, likes to sleep a lot too. "It's like she thinks it's come to life," Campbell remarked. And that's just it—the toys he makes, when welcomed by these children, fold into their lives as companions who share their experiences.

Joyful Effort

As Sir Ken Robinson once observed, "We have evolved this powerful sense of imagination, the ability to bring to mind things that aren't here. And from it follow all kinds of powers like creativity—and uniquely and distinctly—the power of empathy."

Campbell's imagination is a superpower blend of creativity and empathy. Robinson extols this kind of *applied imagination*—the extraordinary and entirely natural capacities of children to make creativity a practical matter, to *do* something with their imaginations. The creatures Campbell makes are the perfect example of what applied imagination makes possible: Campbell gets to enjoy every minute spent doing a craft, and the toys offer sweet distraction and tangible comfort to those who need them most. He loves what he is doing for others. Children offer a profound teaching: When you are engaged in a meaningful task, you are both transported and transformed by it. Your efforts are joyful.

Imaginative Empathy

The power of imagination begins in the cradle of childhood, starting with the very first objects of affection: blankies, teddy bears, dolls, pacifiers. Such objects transport us to zones of comfort; they are part of our earliest acts of creativity, imagination, and pretend play. Psychologist D. W. Winnicott called them *transitional objects*, and any parent or caregiver knows their importance in reducing fears of separation, along with fears more generally. Campbell knows their powers intuitively. The creative connection with objects of affection awaken an underrated aspect of empathy: fantasy. This is our ability to emotionally connect and identify with fictional characters in books, movies, and stories we make up. The link between empathy and fantasy is why two-thirds of children between three and seven years of age have imaginary companions, whether an invisible friend or a personified object.

Children love their "friends," and even as they know the difference between what is real and make-believe, they love and bond in a way that has powerful effects. "Part of what children do with imaginary friends is practice things they are trying to figure out. Young children, especially, don't understand friendship all that well," says developmental psychologist Tracy Gleason. Many presume that children with pretend pals are peculiar kids living in their own worlds. But studies show that the opposite is true, as these children have a slightly larger vocabulary, are quite social, and are good at understanding the perspectives of others. "With imaginary companions, children will talk about the things they do together, or what they both enjoy doing, or how sometimes their companion doesn't want to play. They're trying to understand: What does it mean when somebody doesn't want to play with you? What does it feel like to be rejected? That's a lot of stuff to figure out!"

This doesn't end in early childhood. Teenagers commonly have imagined relationships with celebrities or media figures, called *parasocial relationships*. "Say you're fourteen," says Gleason. "In your mind, you're very close friends with Jennifer Lawrence. Of course, you know it's not real. It's totally one way. You read a lot about her. You watch all her stuff. You get online and you check out all the social media about her. You even go through life thinking, 'What would Jennifer do in this situation?' It's like using her as a sounding board for your own choices."

Imagination even helps adults work through issues, particularly in relationships, because as Gleason affirms, "the essence of being a person is to have connections with other people." A bereaved widow will commonly imagine that his or her deceased spouse is still there. She talks to him in her mind, having conversations about what's happening in her current life, and imagining what the spouse would say. Gleason notes that, while this may be in part due to separation anxiety, after years of living with a person the imaginary bond is based on a very real dynamic. So to maintain it even after a spouse is gone can be adaptive in the short term. More than memories or living in the past, doing so can help the bereaved work through present situations.

Consider Tom Hanks's character in the movie *Cast Away*. His volleyball, Wilson, saved him from loneliness and despair, and preserved his will to live. You can also imagine a kindred spirit for yourself, one you can communicate with and receive guidance from. In this way, imagination makes empathy and kindness, for yourself and others, easier to practice.

Kindness in Practice: *Imagining Friendly Comfort and Guidance*

Think about your life. Do you cling to sacred objects such as photos, rosary beads, jewelry, an old sweatshirt, or treasured mementos that connect you to people you love? These objects of affection fill the imagination and are bound up with sensations, emotions, thoughts, and memories. You can create these associations on purpose by imbuing an object—a stone, feather, or statue, for instance—with comfort, love, and wisdom. Even though you are an adult, you can still use the wonders of imagination to befriend yourself. This is especially helpful when facing life challenges or recovering from disappointments and traumas. We all need a kind and wise friend looking after us, especially one that comes from within. Here is a meditation to connect with one.

> Begin by sitting in a comfortable position with a straight back in a posture of strength, grace, and dignity. Start to relax. Take a few deep breaths, soften any tension, allow your mind to become quiet, and continue with relaxed and natural breathing.
>
> Visualize that you are walking inside a serene place where you feel peaceful, content, and safe. You may prefer a quiet cabin in the woods, a perch on a mountainside, a tree house, or a cave. When you feel ready, invite a kind companion to join you. Let this friend's natural form surface, however it appears. It could be a person from the past such as a grandparent, mentor, or teacher. You could imagine a benevolent religious figure, a deity, an angel, a superhero, a spirit guide, or a power animal. You might simply feel a presence, see a bright light, or feel enveloped in a warm blanket. Allow a wise and friendly being to approach, who knows you well, who has always known you in some way. Feel that this companion cares for you and accepts you completely. You trust and respect your companion. Absorb the feelings of mutual caring.
>
> Bring to mind a question or problem that you are facing. Ask your companion for help. Listen patiently for an answer, allowing your companion to communicate with you in whatever way is natural. Words may come. Perhaps you will sense the answer. Or your companion might guide you on a vivid imaginative adventure full of symbolism. Be open and curious.

Thank your friend, and imagine how you might use your companion's advice. Observe what obstacles might get in the way. Consider what actions would be helpful for you and for others whom you may be concerned about or who are causing you concern or irritation.

Return by visualizing your imaginary haven, and rest in feelings of security and comfort. Know that you can call on your companion whenever you need guidance. In your mind's eye, begin to leave the scene and return to the room you are sitting in. Open your eyes. Recall what was meaningful to you. Take out a journal and record the experience and insights.

Ralph Waldo Emerson said, "Be silly. Be honest. Be kind." Tap the power of your childlike imagination. Use your natural inclination for play and creativity to cultivate tender awareness. Like Campbell's stuffed animals, an imaginary companion can be an object of affection and evoke feelings of goodwill, friendship, and care. At any moment, you can invite your companion to hold your hand, walk at your side, sit nearby, and share advice, comfort, and encouragement. As Campbell said, "I think being kind and not mean will change the world"—and it all starts with your wild imagination.

Reflection: *When I am having a difficult moment, I imagine kindred spirits at my side.*

CHAPTER 7

The Kinship of Belonging to Each Other

Treehouses evoke coziness, playfulness, and safety for me. My best friend, Heidi, whom I met on the first day of kindergarten, had a treehouse that her father built in an old oak tree, twelve feet off the ground. We would huddle up there with books and art supplies, in our own hideaway refuge. Heidi's household offered me more than an awesome treehouse, though. While my parents were separating, her folks bestowed me with an alternative perspective on family life that offered a safe base from which I could explore the wilderness of my childhood. They provided me with true solace, which was a precious gift.

So when a friend told me about the Treehouse Community in the Berkshires of western Massachusetts, an innovative sixty-home development, I was immediately smitten. Turns out I'm not alone in my associations: "A treehouse for me is a happy childhood," founder Judy Cockerton told me. "You climb up into a tree, into this wonderful little safe space, and you look out over the horizon, over the landscape, and you get a whole new perspective."

Cockerton has a very large view. With the Treehouse Community, she wants to reenvision the entire child welfare system. "I wanted to create an intergenerational community to move children out of the foster care system into permanent, loving families: a community that invests in hopes, dreams, and futures every day to change children's life trajectories and outcomes." The village is populated with more than one hundred people: families, children, teenagers, and seniors who help the kids out.

From the moment Cockerton got up the morning I spoke with her, she was pinged with texts from the Treehouse kids. That evening there was to be a holiday fiesta and they were abuzz with preparations.

"I can't wait for tonight!"

"I have my pink dress ready. Do you have any size 8 pink shoes that I could wear with it?"

"Do I have to wear a suit?"

Through the phone, I could hear Cockerton's joy. "They have community, they have friends, they have family, they have a place where they feel safe and belong."

Now that's a pretty cool treehouse.

Kindred Spirits

We need each other. There is great wisdom in having a sense of social responsibility for the welfare of an entire community—from the youngest to the oldest. As Cockerton points out, "It's the way life should be: everyone investing in each other's health and well-being."

My friend Mary Anne agrees. For more than a decade she has been matching the children and teens of incarcerated parents with retired adult mentors. "We've separated our elders and our young people. I think what people are learning is that growth happens when we can put populations together, and we can kind of emerge from there." Every day Mary Anne sees the benefits of the reciprocity of intergenerational mentorship and creating safe spaces for connection. She observes, "There's a lot of mistrust in the beginning. Sometimes the kids test the mentor, like, 'Are you going to come back?' Or they'll act out a little bit. There are so many young people, even at age seven, who think they're not liked or that there's something wrong with them. Sometimes it's just simple conversations, showing up, or things that we do with one another that show kindness. It doesn't have to be the big massive things."

This yearning for communal caring is deeply embedded in us. Once upon a time children were held, carried, and passed from one person to another as groups traveled across savannas, deserts, and mountains. Wrapped in slings or papooses, carried on shoulders, or held by the hand, it was high-touch caregiving, shared across genders and generations, involving mothers, fathers, aunts, uncles, grandparents, siblings, and near kin. This yearning to come home to ourselves and each other is what ecophilosopher Joanna Macy describes as "intimate mutual belonging."

Kinfolk. Kinsmen. Kindred. Kinship. "Kin" means "sameness," "relatedness," and "of the same kind," and it evokes feelings of closeness and friendliness toward others. At its root, the word "kindness" is aligned with kinship and is about community.

In-Person Social Networks

Whether your distant ancestors lived in caves, tree huts, or tents, it's clear to evolutionary historians that humans adapted to environments and survived by helping each other and forming alliances. Along the slow timeline of humanity, the most caring members of a community—the ones who ensured that kin and children survived—carried these caregiving genes forward. Natural selection has favored prosocial traits like empathy, kindness, sharing, cooperative play, mutual understanding, perspective taking, and trust.

Primatologist and evolutionary theorist Susan Blaffer Hrdy proposes that humans thrived as a direct result of cooperative breeding in tight-knit communities of helpers. She attributes this advantage to *allomothering* or *alloparenting*, in which caregivers other than the biological mother or father assume caring roles. We find strength in numbers. Judy Cockerton is proving this through her successes, and she has four more Treehouse Communities in the works.

At the same time, living in community can feel hard. Many people suffer stressors within the family crucible, and the more stressful our experiences in childhood are, the greater the impact on physical and mental health. The landmark Adverse Childhood Experiences (ACE) Study was conducted twenty years ago. Its simple, ten-item questionnaire covered heartbreaking experiences including abuse, neglect, and family dysfunction. The study found that, among more than seventeen thousand people surveyed:

- 64 percent of adults reported at least one adverse childhood event before the age of eighteen, and 12.4 percent reported four or more ACEs,
- 28 percent of adults reported physical abuse,
- 21 percent reported sexual abuse, and
- 19 percent grew up with a family member with mental illness.

ACEs are very common. Almost two-thirds of Americans have an ACE score of 1 out of 10 events, and many of us (myself included) have more. However, those with a score indicating 4 events or more are at greatest risk for adult physical and mental health problems. The science is unequivocal about their effects: they trigger a chronic stress response in the body, alter immune function, and affect brain chemistry. The study compelled pediatrician Nadine Burke Harris to sound a rallying cry to screen all children across the nation for ACEs at their regular physicals. "Children are especially sensitive to this repeated stress activation, because their brains and bodies are just developing," says Harris. "High doses of adversity not only affect brain structure and function, they affect the developing immune system, developing hormonal systems, and even the way DNA is read and transcribed.

"If my patient has an ACE score of 4, she's two-and-a-half times as likely to develop hepatitis or chronic obstructiove pulmonary disease, four-and-a-half times as likely to become depressed, and twelve times as likely to attempt to take her own life—in comparison to a patient with zero ACEs."

With awareness of a problem, treatment becomes possible. One of the study's cofounders, Robert Anda, says that what is predictable is preventable. Having a high ACE score does not mean a life sentence of misery. A score can be used to design compassionate responses such as prevention services, training in resiliency skills, self-compassion exercises, programs that help break intergenerational cycles of stress, and using trauma-informed policies and programs in communities and schools. Foster children represent the most vulnerable populations, which compelled Judy Cockerton to respond compassionately and to provide them with safe spaces. To Cockerton, it comes down to offering basic needs. Safety. Love. Belonging. And throw in some pink dance shoes.

Kindness in Practice: *Creating Safe Places*

No matter where you are in the trajectory of your life, forming kinship bonds with others and feeling safe are foundational. You may be wondering about your own ACE score. You can easily find the questionnaire online and discuss the results with a health provider.

The following exercise will help you notice when and how you have found solace and safety throughout your life.

> Start by reflecting on the moments in your life when you have felt safe and cared for. What was that like? Who supported you? Were there people you could trust? Maybe there was a person who made you feel loved and safe, who always had your back or offered the perspective on life that the world can be a friendly place. What place could comfort you? Maybe it was a childhood playhouse or a nook in your closet or under your bed. Have there been activities that soothe? Perhaps coloring with crayons or listening to a certain type of music?
>
> Grab your journal. With kind awareness and tenderness, complete this sentence:
>
> *I feel a sense of comfort and safety when I remember…*

A few years ago, I wrote a letter to Heidi's mom to thank her for being such a beneficial presence in my early life—one of several *allomoms* I had found among neighbors. Mrs. Mel is now in her late seventies, and I had not seen her in decades. I was surprised when she wrote back, saying she wished that she had been more aware or had offered more support. But that was just the thing: her natural caring was all I needed.

We can intentionally create models for communities like Treehouse Community, and we can also bring together the multiple generations in our neighborhoods to show the world what can be done when we remember that we belong to each other. As Cockerton contemplated, "Kindness is blessing someone's life."

Reflection: *To be kind to others is to feel kinship with them; when I do this, I can bless someone's life.*

PART 2

Your Caring Mind

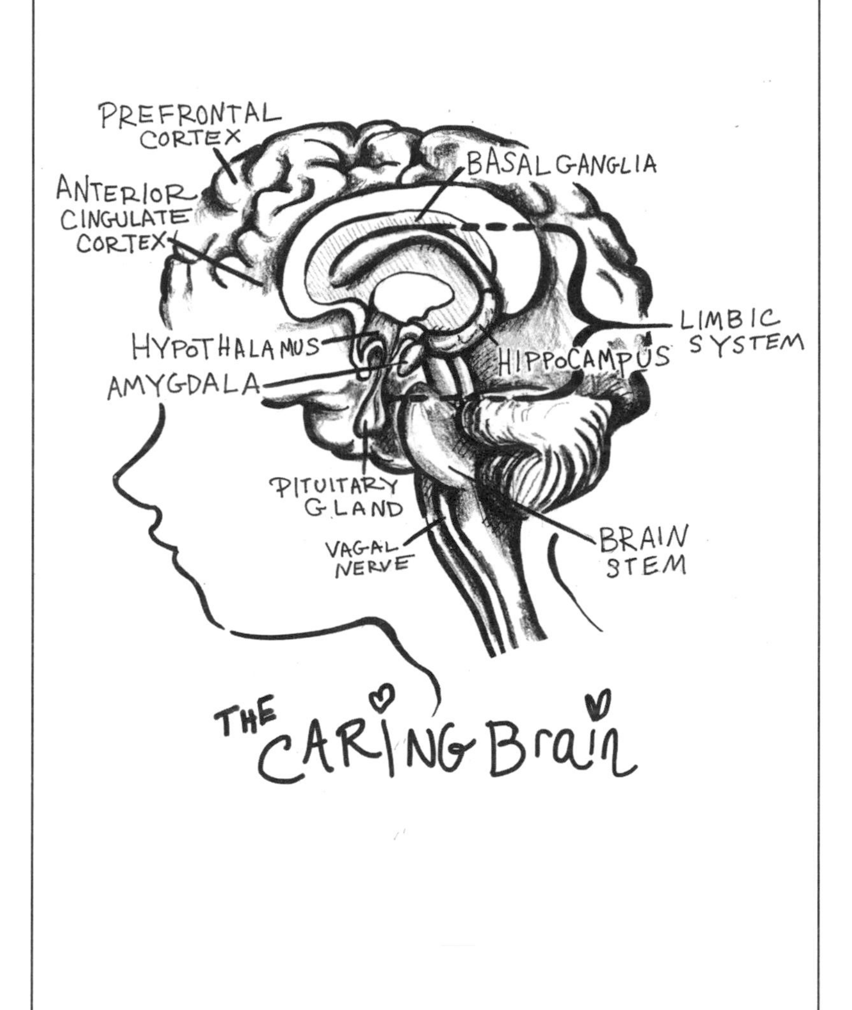
PREFRONTAL CORTEX
BASAL GANGLIA
ANTERIOR CINGULATE CORTEX
LIMBIC SYSTEM
HYPOTHALAMUS
HIPPOCAMPUS
AMYGDALA
PITUITARY GLAND
VAGAL NERVE
BRAIN STEM
THE CARING Brain

CHAPTER 8

Reset Your Stress

"When we come back from the war zone and get out of the military," explains veteran Justin Blaze, "we immediately lose camaraderie, we lose our community, we lose our family. Nobody understands us. That's why veterans become so isolated and there are so many suicides in the United States." And that's why Blaze created a yoga curriculum used nationally for veterans and their families. "A free community class essentially opens a door to being with others like us. Veterans see that there's no judgment, that other people who are like them are doing yoga, so something must be working. That's the only reason they come." And along the way, they get the same results Blaze did.

"Usually a soldier does one, two, or three tours, and that's more than enough. I did more than forty trips to Iraq and Afghanistan in ten years" says Blaze in a mild-mannered way. "The stress added up. In the military, you are trained to operate above the normal optimal zone for stress. When everything in your body is telling you to run away, you are trained to stop and turn and go toward it. It's hard to turn that off." Over time, Blaze accumulated many injuries and a lot of battle stress. He was irritable, mean, and just didn't feel like himself.

Then Blaze's roommate invited him to try a yoga class. "It was very challenging physically. That hooked me. Because it was low-impact, given my injuries, I could do it. Then at the end of class, when I was just laying in *Savasana*, corpse pose, I felt a sense of rest and relaxation I hadn't felt in years. I got a little taste of it and I wanted more." For Blaze, learning how to "reset" himself physically and mentally was exactly what he needed after years of being on a high-adrenaline rush. He began to improve in both body and mind, and he found himself being kinder and gentler with himself and with others. He became a certified yoga instructor.

"I felt obligated to give back to people. If there are other people who are even barely dealing with the things I was dealing with, I have a need to help them." That's why he started his nonprofit outreach program, VEToga, which provides yoga classes and teacher trainings for veterans and their families.

Your Internal Alarm System

Chronic states of stress lead to wear and tear in the body, and possibly to more serious illness. The mental and emotional effects of chronic stress can be painful too. It's hard to focus and concentrate, and you might even get mad at yourself about not feeling quite up for life's demands. Moreover, it is harder to feel kind or compassionate.

This is because when your internal alarm system is in survival mode and the switch is stuck so that the system's always on, you're too busy reacting to stress to respond with balanced perspectives. You can't call on the higher regions in the neocortex of the brain, such as the *prefrontal cortex* (PFC), which enables you to be aware of your own thinking, keep perspective, plan, make sense of your social world, and feel compassion.

Here's what happens. The earliest living creatures relied on quick, short-term stress responses to predators—otherwise, the consequence was death. This survival instinct is found in the *limbic system*—which includes the amygdala, hypothalamus, basal ganglia, and hippocampus. Your limbic system is always on, alert and scanning the environment, whether you know it or not. When you encounter something that feels threatening, the limbic system trips the alarm system switch, and you manage stress as quickly as possible by fighting or fleeing—or even fainting. Once the alarm is triggered, the hippocampus sears your reactions into memories and autobiographical stories, lest you forget the danger. But this can also scar you, which is why you vividly remember the most traumatic things that happened in childhood and why experiences like Blaze's remain so potent even after they end. As he said, "I was stuck in fight or flight, there's no question."

You may be struggling with some of these experiences yourself just navigating a stressful and busy world. Or you may be struggling with *inner* threats or negative habits of mind. We all cycle through unhelpful or critical thoughts, which also cause distress—whether we are aware of them or not. That's because our limbic brain also interprets harmful inner conditions as threatening. You may be flooded by negative world news, worry about paying college tuition for your children, have a work deadline you don't know how you'll make, or harbor persistent self-doubt about your body, love life, or career. The way the limbic system encourages you to respond to stress is adaptive in the short term, but when you're constantly on high alert, there's no chance to recover from one stressful experience before engaging another. High levels of stress hormones have serious physical and mental health repercussions. You simply aren't built for prolonged stress.

This constant stress response contributes to falling under a SPEL, short-circuiting motivational empathy and kindness, as described in chapter 1. Blaze points out that "When you're in fight or flight, you're in a desperate state. I saw myself being irritable, being reactive to people, being mean. That wasn't who I was, and it was upsetting." He needed to learn how to rest and restore his equilibrium. He had to retrain his brain.

You Can Reset Your Stress

If you are living in a state of chronic stress, you are not alone. The American Psychological Association reports that stress levels are at an unprecedented high. Even so, there is a lot you can do to remedy it without overhauling your lifestyle, and it begins by noticing the effects of stress and relaxation on your body.

Think about the sensations you feel when you do something stressful, like giving a public speech, going on a first date, or reading an explosive text message. A cascading reaction instantly unfolds that signals the release of stress hormones. You might experience this as a racing heart, dilated pupils, immune suppression, slowed digestion, needing to pee, tunnel vision, and hearing loss or numbness. One way or another, your body is working to get you back into a balanced state of being. This happens through the *autonomic nervous system*, which is rigged with two branches: The *sympathetic nervous system* energizes you for action, preparing you to react quickly to fight or flee. The *parasympathetic nervous system* works to calm you back down. This "calm and connect" process regulates various things including your breathing and heartbeat, and it triggers neurochemicals—dopamine, oxytocin, and vasopressin among them—that work to quiet down the nerves. But in some extreme situations a person's parasympathetic response may begin to shut down, triggering the freeze-or-faint responses. Depending on the situation, the autonomic nervous system can be energizing and motivating, or it can immobilize you in the effort of self-preservation.

For the most part, most of us do not face extreme stress or battlefields but rather the daily hassles of living. We can learn to balance out the typical stress response by intentionally engaging our internal calm and connect system. It soothes you, and it aids resting, processing experience, and refreshing—physically, mentally, and emotionally. You can encourage it. As Blaze points out, "Coping mechanisms like yoga, meditation, mindfulness, walking in nature, spending time with your dog, and spending time with loved ones all engage your parasympathetic nervous system. It offers a sweetness that you can't ignore. You don't know what it is, but it tastes good in every way, it makes you feel good in every way, and you feel calmer because of it."

When your bodily systems are in rhythm, and you've cared for yourself by attending to both of these basic states, you become more capable of being readily available for others, to "tend and befriend" them. Tending and befriending is also an adaptive response of the calm and connect system, whereby we instinctually protect, care for, or quiet children and loved ones in the face of threatening situations, or seek the safety of others in our social group. We can also intentionally befriend ourselves through self-compassion, as we shall see in chapter 10. Being aware and skillful about managing your own stress breaks the SPEL and naturally makes you a kinder person.

Kindness in Practice: *Your Stress Style*

You can learn how to respond to stress in beneficial ways by first identifying what the stress response feels like in you. This way, you can become aware of a reaction at the earliest signs. Here's how:

> Choose a recent upset or stressful event, something that was mildly or moderately disruptive. In your journal, draw a stick figure or an outline of your body like the one pictured here. Write or draw in any sensations that you experienced during your upset, or any that

occur in the moment as you recall the experience. Your style of stress is unique to you. You may clench your jaw, feel tightness in your shoulders, get hot, have a hard time falling asleep, or feel confused. Maybe it becomes a quick or slow-brewing temper, digestive discomfort, back or neck pain, or social isolation. Here are some prompts that you can use as you reflect on stress's many effects. What is your stress style?

If stress were a color, it would be…

The picture that comes to mind with the word "stress" is…

My stress symptoms include…

I feel stress in these parts of my body…

I know I am stressed when I emotionally feel…

The very first sign of stress is…

When I'm stressed, my thinking becomes…

Others can tell when I am stressed because I…

When I'm stressed, I become the kind of person who…

Once you have your description, during the week that follows notice when you begin to feel signs of stress. How does simply noticing this affect you?

Now that you know how your body's alarm system works, and can identify how you experience stress, you are empowered. As some say, "What you *feel*, you can *heal*." This is what hooked Justin Blaze on yoga and mindfulness practices; they countered his stress and restored his mind and body. It has proven just as beneficial for other veterans with PTSD, a benefit that led him to ask, "How can I share the gift that I've been given with others, who then share it with as many people as possible?" Blaze broke the SPEL when he realized that his well-being can make a large impact on his family, community, and everybody around him. You can too.

Reflection: *While I'm wired for stress, I'm also wired to relax. What can I cultivate in my life to feel more at ease?*

CHAPTER 9

Befriending Your Senses

On the night of a rare celestial event—a lunar eclipse with a supermoon that was to glow red—my daughter Josie and I drove up a hillside to watch it. To mark the occasion, we each wrote down our intentions and set the pages aflame to release our wishes to powers bigger than us. Josie and I then leaned against the car and stared at the sky, waiting for Earth's shadow to slowly mask the moon.

"What did you write?" I asked her.

"I'm not telling."

"Okay, fair enough." We lit the rest of the matches for the fun of it, thought we heard a coyote in the distance, and tried to spot constellations through the city-lit sky. We took it all in, with every sensory system in full receptive mode.

After a while, Josie wondered, "Doesn't it seem that we are just so small, like we can't make any difference in the big scheme of things?"

I countered, "Doesn't it seem like we are completely connected to the entire universe, every bit of us, and everything we do matters?"

I will always remember that night. The sound of crickets. The scent of pine. A warm evening breeze. A starlit sky. A penny-colored moon. My daughter and I, shoulder to shoulder, sitting on the hood of the car asking big questions. It was a reminder that we are sentient beings, fully awake and alive.

Senses as Signals

As Rumi wrote, "The body is the house in which the spirit resides." When we stop thinking long enough to reconnect to our bodies, we discover the spirit that lives there. Your body is a sensitive instrument capable of responding skillfully to inner and outer perception. The word *sentient* implies your ability to feel, see, hear, smell, touch, and taste—to be aware. These senses are the foundation of humanity. By honoring the power of this natural sensitivity, you become more aware of the majesty of your life and just how much it matters. You become empowered to guide it.

Meditation teacher Tara Brach writes, "Presence is the awareness that is intrinsic to our nature. It is immediate and embodied, perceived through our senses." With mindful awareness, you can experience sensations for what they are: signals that invite you into the present moment. Sensations are the language of the body. Rather than engaging an ongoing, mental dialogue that keeps you locked into the past or takes you far into an uncertain future, you can simply be. Instead of indulging chains of reactivity that may prolong or expand your suffering, you can return to your spirit's house.

By attending to sensations, you gain a skillful way of becoming present in your own body before interpreting and responding to the world. Brach describes three qualities of *presence*:

- Moment-to-moment wakefulness,
- Openness to the flow of life, and
- Responding to life's joys and sorrows with tenderness.

This embodied intelligence inspires inner wisdom, intuition, compassion, creativity, and synchronicity with others. So befriend your senses with curiosity and wonder.

Your Sixth Sense: Awareness

In busy and stressful times, you risk tuning out in order to cope. As you narrowly focus on what needs to be managed in the moment, you tune out the very experiences that can make you feel joyful, wholehearted, and connected. These are often visceral experiences that draw on your senses as you look at a poster on the wall, listen to your teenager's tone, smell coffee before you drink it, taste the freshness of toothpaste, or touch the blanket you curl under. Your senses provide you with a constant stream of information and, whether you realize it or not, you reflexively assess each piece of information as pleasant, unpleasant, or neutral. This determines how you remember the past, experience the present, and move into the future. Your senses are therefore a key not only to surviving but to thriving.

You have a sensor system that recognizes bodily sensations and even the physical effects of your imagination. This awareness is called *interoception*, and it happens both unconsciously and consciously. Steven Porges describes interoception as a "sixth primary sense." Its main component is your *vagus nerve*, which forms sensory and motor pathways from your brain to internal organs. Being attuned to bodily sensations with gentleness—simply observing the signals—increases interoception.

Doing this without evaluation or judgment is important. Some people can become hyperaware of internal body states and overidentify with sensations, which leads to anxiety or panic, sets the limbic system in overdrive, and makes it hard to engage your calm and connect system. As trauma expert Bessel van der Kolk says, "People who cannot comfortably notice what is going on inside become vulnerable to responding to any sensory shift by either shutting down or by going into a panic—they develop a fear of fear itself." When we lack the ability to witness sensations, van der Kolk points out, we end up seized by them.

Deliberately witnessing our senses with curiosity, kindness, and tenderness allows us to harness our natural preference for good experiences such as emotional balance and resilience. We find that being present as life unfolds, without reactivity or judgment, without feeling confused or overcome by difficult feelings and sensations, instills a comforting sense of security. And we come to appreciate the kindness this presence evokes, for ourselves and for others.

Kindness in Practice: *Taking in the Senses*

At times, you may feel like a time traveler caught in thoughts about the past or worries about the future. You might forget to tune in to your body, a beautiful instrument that can bring you back to the present moment and communicate about what's going on in yourself and your surroundings. Being in the present moment starts with noticing what you sense right now. Learning interoceptive skills and expanding your body awareness takes time and practice, and it may be helpful to seek the guidance of a teacher, counselor, or friend. This meditation is one of many ways to connect with your sensory intelligence.

> Take a few moments to sense your surroundings. Feel the ground supporting you. Begin to tune in to your senses: sight, sound, smell, taste, and touch.
>
> Wherever you are right now, begin to look around. What do you notice about your environment? Notice some details: light... shadows...colors...shapes...movement...texture.
>
> Notice the sounds around you...behind you...in front of you...all around you. Hear the different tones and rhythms. Perhaps there is a breeze or the hum of traffic.

Maybe you notice a smell; maybe it is fragrant, evocative, strong, or even unpleasant. No matter. Simply notice the sensation of smell.

Perhaps you are chewing gum or eating. If you are, notice the flavor and the texture. Maybe you are experiencing something new about the taste. Just notice what that may be.

Now notice yourself in your world. Imagine yourself as a friendly observer who is watching you; notice how your body is right now. What are you feeling? Notice the sensations in your body: tension, discomfort, vibrations, tickles, or subtle movements.

Feel your feet on the floor or grass; notice the texture of clothes on your body. From this feeling of being grounded, tune in to both inner and outer realities by noticing what's happening *inside* as you sense the world *outside*. How do they relate? Simply be aware.

Take in a relaxing, deep breath and soak up all these sensations from a variety of sources. Notice how you are part of a very alive and connected world.

Over time, perhaps because of adverse childhood experiences (ACEs), you may have learned to split your mind from your body to cope, which leaves you disconnected from the world around you. By recognizing that your senses are helpful signals, you can take back control and choice in your experience of life. You are more equipped to move forward peacefully and deliberately. When you want to be soothed or inspired, all you need is something embodied and sensually stimulating, like watching the moon rise, listening to waves crash, baking bread, eating an apple, or wearing a soft sweater. As simple as it is, this connects you to your body, to all beings, and to the whole world.

Reflection: *My body is a wise messenger. As I befriend my sensations, I am aware of the majesty of my life and just how much a part of the world I am.*

CHAPTER 10

Emotional Paradox

I know the exact moment I decided to become a therapist. I was eleven years old. For a long time, my life had felt fraught with anxiety, a temper, and an expectation of impending doom. There had been days when I refused to go to school. I wanted to stay home to protect my mom, because there wasn't much trouble when I was there. I had been taken to see Dr. Moe, the guidance counselor, a lot. Those meetings were excruciating. Defiantly, I refused to speak, and I managed to survive elementary school in stoic silence. Because I had a hard time focusing, I was kept back in math.

In the middle of the sixth grade, my parents had split up and my dad finally left town. I was promoted to a high-level math class, which was no small achievement and reflected how much better I was feeling about life. It was the tentative budding of self-confidence. On the first day of my new class, Mrs. Dulfer was handing out papers with a scowl on her face. As she walked along the rows of desks, I realized she was handing out a test and I started panicking. Timidly, I raised my hand.

Mrs. Dulfer glared at me, "*What* is it that you need?"

"Um, do I have to take this test? I haven't been in this class before," I whispered.

Then she said, "What did you say, Dearie? I can't hear you."

I raised my hand even higher, courageously squeaking out, "I don't know what's on this test. Do I have to take it today?" Once again, she implored me to repeat myself. I did. Then Mrs. Dulfer waddled over to me, grabbed the top of my hair, and pulled me straight up. Stretched to my tippy toes, I watched her mouth hiss at me:

"Don't you *ever* raise your voice to me again."

Dangling by the hair, as much as my scalp burned, the red-hot shame I felt in front of the class was infinitely worse. She let go and, as soon as I found my footing, I ran out of the room and down the hallway for refuge with Dr. Moe. His office door was open, so I rushed in and collapsed on the chair in front of him. He asked, "What's the matter?"

With tears streaming down my cheeks, rubbing my scalp, and nearly hyperventilating, I sputtered out more words than I had ever spoken to him at once. I told Dr. Moe what happened.

There was a long, considered pause. Then Dr. Moe looked at me and said, "Tara, are you *sure* you didn't raise your voice?"

WHAT? My heart stopped. Could it be that Dr. Moe wasn't on my side after all? I realized, then and there, that every adult had failed me in some way. I felt profoundly misunderstood. I glared at Dr. Moe and I thought to myself, *I am going to help kids like me and actually listen to them. I am going to do a better job than you!*

Cultivating Mindful Compassion

Failures of empathy occur frequently. Most of the time they are unintentional or accidental—little slights or simple misunderstandings. Other times they can be more severe: outright discrimination or blatant mistreatment. That moment with Dr. Moe is etched into my memory, especially because of the mixed emotions and the flurry of thoughts that crashed together within me. "Our emotions are the sources of our most meaningful experiences in life, but they can also lie at the root of our deepest problems," write psychologist Paul Gilbert and former monk Choden. In sixth grade, what I didn't realize was that I had tapped a *compassionate motive*, a desire to alleviate suffering paired with an inkling that I could do something about it.

A Clash of Minds

Mindful compassion "is about recognizing the benefits for deliberately harnessing our caring motives as a way to organize our mind," write Gilbert and Choden. That's why understanding how your emotional brain works is essential. When you get angry or upset at someone or something, or beat yourself up with self-doubt, you may actually be of two minds and not even know it. An internal conflict may arise. *Is there something wrong with me or with this situation? Am I open to a new adventure or do I play it safe? Am I being selfish or nurturing of myself? Do I accept or reject his or her affection? Am I being kind or cruel if I stick to my boundaries?* These kinds of mental conflicts can help you understand why you get caught up in unkind mindsets, manners, or misdeeds.

Paul Gilbert attributes the paradoxes you can experience in your own mind to the encounter of what he calls the "old brain/mind" and the "new brain/mind." Your "old brain/mind"—which we've been exploring in the limbic system—is the "base model" of human emotional regulation and hasn't changed much over millennia. Its job is to serve your basic survival instincts as soon as possible and to seek out pleasure and comfort. It is speedy and reactive. Gilbert describes the three main systems operating within it:

- **A threat and self-protection system** that senses threats quickly and activates the fight-flight-freeze-faint response in your limbic system. This is like your home surveillance system.
- **An incentive and resource-seeking system** that propels you to seek pleasure, consume, play, and mate. It's like an Energizer Bunny scurrying about, looking for fun.

- **A soothing and contentment system** that seeks balance, rest, and connection, and is strongly linked to affection, bonding, caregiving, kindness, and compassion. This is the calm and connect system described earlier, and it is a bit slower to come online, but when it does, it gives you a sense of overall well-being—like a baby's snuggly or a rocking chair.

Your "new brain/mind" developed later in human evolution. It's really smart. The newer model is more complex and allows you to work things through, compare, contemplate, mull things over, create, innovate, imagine, seek knowledge, strive for goals, and develop an identity. This allows for quick learning, exchanging information from among groups, and passing on these adaptive genes to future generations. Importantly, this sophisticated upgrade allows you to be *aware* that you exist and have a sense of self. Thanks to your "new brain/mind," you can *be aware of your awareness*, unlike any other animal, and observe your own mind. This is, of course, both a blessing and a curse.

When your "new brain/mind" is pulled by the fears and passions of the "old brain/mind," you can get stuck in unkind behaviors. This is the unfortunate bug in the system, so to speak. Psychologist Rick Hanson warns about a *negativity bias*, a quirky inclination to scan for danger, to interpret things in unhelpful or harmful ways, and to remember unpleasant events. This leads to habits like rumination, engaging in negative social comparison, doing harmful things, and fixating on unhelpful or undesirable thoughts, worries, memories, and situations. As Hanson likes to say, your brain is more likely to stick to negative events like Velcro, while it lets positive events slide off like Teflon. A root of most human misfortunes and calamities can be linked to the primitive drives to survive and seek pleasure over pain as well as to the fear-based, emotional reactions that get provoked when faced with threats.

It's not that the base-model brain is bad and the new version is good. You need them both for your mind to interpret the world. But there are competing needs and desires at play. In modern life, these can manifest as:

- **An internal conflict**: "I want to feel close to you, but I'm afraid you'll abandon me, so I'm breaking up with you."
- **An inner critic who wreaks havoc on your self-esteem**: "I'm not good enough, so I won't bother applying for the job."
- **Engaging in unhealthy habits**: "I know I need to watch my blood sugar, but just one cookie won't hurt."
- **Choosing harmful behaviors to cope with fear, anxiety, or shame**: "I'll have another drink, since no one will notice if I'm home late."

When you understand how your brain's old and new models can work together or in opposition, you realize that you have more control over your experience than you previously thought. You can be *intentional* about life. What's more, our human capacities to read each other and be curious about the experiences of others mean we can be intentionally compassionate. In the words of Gilbert and Choden, "We have the ability to empathize and imagine what it's like to be another person; so the smart human qualities of our minds—and they are special—can be put at the service of either harming other people or helping them." Of course, this also applies to how we treat ourselves.

Kindness in Practice: *Bringing Clarity to a Headspin*

You can easily get caught in a headspin, a never-ending loop of negativity, because the competing needs of your older and newer brain models can really trip things up in your mind. For now, begin to notice how this works in your experience. Reflect on how your mind can at times be your greatest advocate and at others be your worst enemy. You can't help but react in one way or another to your inner and outer worlds. Without judgment or scrutiny, journal responses to the following questions:

When do you feel critical of yourself? Of others?

Sample responses: When I can't accomplish my to-do list. When I struggle to pay the bills. When my son does something before thinking of the consequences for the family. When I watch the news.

When do you feel kind toward yourself? Toward others?

Sample responses: When I am knitting a sweater for someone. When I remember how much work it took to get where I am in life. When my kids do something nice for a neighbor. When I read about something good a nonprofit organization is doing.

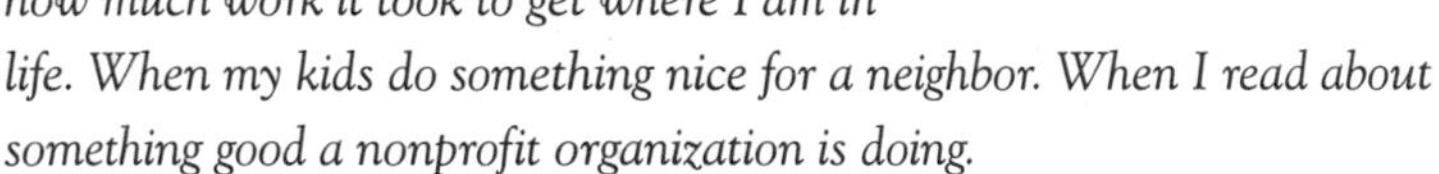

Six years after my fateful math class, Dr. Moe was transferred to the high school guidance office. He wrote me a college recommendation letter. I was shocked when I read it. He said he was impressed by what a hardworking and kind person I was, and that after knowing me for many years, he was confident I could overcome whatever obstacles came my way. He said I would succeed at whatever I put my mind and effort into doing. Surely, he could have said that about anyone. But he was right about me. Because the same mind that experienced red-hot shame, the devastation of being misunderstood, and fury that another person had failed me, turned toward helping others. It inspired me to become a compassionate therapist—and nothing got in my way.

In the same way, you can notice the negativity bias in you and resist it. You can direct that energy into a new resolve: to be kinder than those who hurt you, to support those who are struggling as you once did, and to give others the things you once desired. Because while your mind may feel messy, conflicted, uncertain, and full of paradoxical urges, you can simply be still, observe, and then choose a kinder response.

Reflection: *When I notice paradoxes of mind and heart, I am more intentional about how I respond. I choose to cultivate positive and compassionate experiences.*

CHAPTER 11

The Power of a Pause

One day, one of my favorite yoga teachers, Hania, began a class with a personal parable. "Yesterday, I had a busy day running errands, trying to get everything done before picking my sons up from school. I finished at the bank and, as I stepped outside the door, I saw a car pull up next to mine. The car door opened so far that it pressed against the side of my brand-new car, which I worked so hard for." Everyone in the class nodded or groaned, empathizing with how protective we feel of pristine car exteriors or something new.

"A wave of panic swept through me. My first instinct was to yell at the man, possibly curse him out, and create a scene. But instead, for some reason, I paused and just watched the man slowly get out of his car." Hania pantomimed his sideways lean, with a shoulder pushing against an imaginary door. "I saw that he only had one arm. Then I scanned his car. I saw a veteran sticker. In that split second, I thought about all he may have endured during his military service. My heart stopped for a moment and I felt overcome with emotion, going from anger to compassion. I gathered myself and held the bank door open for him. He thanked me and went inside.

"When I got into my car and started to drive away, it occurred to me that I felt really good. And I even forgot to check the door for a scratch. It didn't matter anymore."

A New Perspective

Hania's story highlights the power of a pause in everyday life, a micromoment that invites kindness. There's a reason why nuggets of advice like "think before you speak," "chill out," and "catch your breath" persist. A pause can make a big difference. Without a pause, you might react instinctually and protectively to avoid the threat or lash out at it. Oftentimes the instantaneous reaction is unpleasant because it is physiological and beyond your control. But with a pause you can skillfully slow down and give yourself enough headspace to consider the situation at hand. After a pause you can respond with clarity and empathy. You can give yourself room to gain perspective.

Pausing allowed Hania to activate her brain's compassion network. As a yoga teacher intentionally practicing mindfulness and as a mom endlessly cultivating patience, she was able to witness her knee-jerk reaction: the urge to yell and curse at the man. And to stop herself long enough to take in the entire scene. When she did, empathy led her to a more understanding response, kind action, and positive feelings that lifted her spirits.

The Prefrontal Cortex: Your Compassion Coach

When you think someone is being unfair, is different from you, or appears threatening, it can feel easier to leap to conclusions rather than engage your natural capacity for empathy. But when you can gain perspective, empathy arises and then transforms into kind action.

The *cerebral cortex* is the most recent evolutionary development in your beautiful "new brain/mind," and it is like command central, where complex mental processes or "executive functions" take place. It sits behind the forehead and includes the *prefrontal cortex* (PFC), the part of the brain that enables you to think in intricate ways. Social neuroscientists are finding that the medial PFC, in particular, is involved in empathy, emotional regulation, self-conscious emotion, and perspective taking. It also communicates with a very important structure called the *anterior cingulate cortex* (ACC), which bridges the limbic region and the cortex—linking sensations, emotions, attention, and social awareness. When the limbic system is triggered by stressors or a threat, the ACC pathway deactivates, and that's the point at which we lose access to our higher, executive functions in the cortex. This is why the ACC is an essential contributor to your ability to regulate your emotions, connect to yourself and others, read social situations, and, therefore, generate kindness and compassion.

Your empathy networks reside across several neural networks in the brain, and new discoveries suggest important nuances that can lead you in different behavioral directions. Empathy, as discussed in chapter 3, allows you to resonate with the feelings of others. When you have an empathic response to someone's suffering, you experience one of two reactions:

- **Empathic distress.** One reaction is to get so upset that you lose yourself in the painful situation. Neuroscientists are discovering that empathic distress triggers brain networks associated with responses to physical pain. These areas encompass the *anterior insula* and *anterior middle cingulate cortex*. You are likely to experience negative emotions—such as fear, anger, disgust, or aversion—and have an impulse to withdraw. We turn away from others, avoid, or numb ourselves. This is how, over time, caregivers and helping professionals burn out.
- **Empathic concern or compassion.** The second response is a desire to help. You respond to an upsetting person or situation with positive emotions such as warmth, connection, and caring. You are more likely to approach others with concern and be more prosocial. This compassion-related brain network encompasses the *medial orbitofrontal cortex* and *ventral striatum*, which are associated with feelings of love, affiliation, and pleasure.

The artful goal is to intentionally move from a state of empathic distress to a state of compassion. This takes some emotional muscle. Social neuroscientists Tania Singer and Olga Klimecki point out that there are factors affecting our empathic responses, including gender, group membership, familial conditioning, and perceived fairness of another person or situation. But the empathic concern network can be strengthened through meditation practice and compassion skills. Singer's team has been putting meditation to the test and, in the process, revealing the compassionate brain's *neuroplasticity*, which is its ability to form new neural connections in response to changes in life situations or environment. Short-term compassion training over several days promotes positive states, even when people are exposed to film clips of other's suffering. But we don't need to be in a lab to test this for ourselves. We can practice small doses of compassion in our daily lives and see how our moods can shift and lift.

In another study, people were given nine months of mental training that involved practices to develop awareness of the present moment; training in compassion, self-kindness, and emotional regulation; and perspective-taking skills. The training positively affected the brain's neuroplasticity. And it caused a thickening in the brain's *gray matter*, which processes information in ways that affect sensory perception, muscle control, speech, memory, emotions, decision making, and self-control.

By doing these kinds of practices, you strengthen your PFC and can more easily understand other people's lives and build the inner strengths—such as resilience, generosity, forgiveness, and compassion—that can be acted out through expressions of kindness. The more we practice, the better we get at responding to life's hassles. Hania has a favorite mantra she likes to say in her yoga classes: "We can get through anything, one breath at a time." The pauses add up.

Kindness in Practice: *Using Your Breath to Pause and Connect*

You can train in the fine art of pausing. Often, all you need is your breath. Tara Brach refers to this as the Sacred Pause. When you breathe slowly and deeply, a new moment is born, because the breath both calms and inspires.

One of many definitions for breath is *spirit:* the force within a person that lends the body life, energy, and power. You can use your breath to find light in a dark tunnel by igniting your caring circuitry anytime and in any place. Taking three deep breaths is a portal to being present. It's one of the kindest things you can do for yourself and for others. As you do so, your body posture, such as a leaning in with a slight tilt of head, can enhance feelings of compassion. Here is a simple way to train yourself to pause, intentionally evoke your body's ability to calm, and then emerge ready for a bigger perspective.

Find a comfortable place to sit or lie down, feeling your body supported by the floor or a chair. You can also stand quietly supported by a wall. Place your hands on your lower belly, on your heart, in your lap, or at your sides. Close your eyes, if you like.

Take deep breaths to fill your diaphragm. Breathe in, down to the bottom of your lungs, and notice your belly begin to expand. Exhale and release the flow of air. Repeat.

You will feel your belly rise and fall with each breath. Feel the flow of air through your nose and down into the bottom of the lungs. Breathe in a natural and gentle way, the way a sleeping baby or a contented cat does.

Follow the rhythm of your breath, noticing your body rise and fall. Let yourself feel the inner spaciousness of presence. Listen to your moment-to-moment experience. Turn toward yourself with kind awareness. You may imagine yourself being held or wrapped in a warm blanket.

As you breathe in and out, you may find it helpful to quietly say the word "peace" on the inhale and the word "calm" on the exhale. Or simply say something like, "I am okay," "All is well," or "I hold myself and others with tenderness."

Do this breathing exercise as often as you like. Gradually you may want to try this for five, ten, or more minutes at a time, expanding your ability to pause.

Once you become used to working with your breath, try adding a simple compassion breathing technique that helps in difficult moments. This one is a variation of the Buddhist *tonglen* practice, which is a giving and receiving meditation practice. As Rabbi Rami Shapiro writes in *The Sacred Art of Loving Kindness*, "It is a way to take upon oneself the pain of the world and transform it into love." Here is a simple version.

> Sit quietly and comfortably, perhaps with a hand on your heart. Breathe in and out in a comfortable way.
>
> As you breathe, bring to mind a sense of warmth, comfort, and ease, or whatever you need in the moment. Inhale this soothing feeling.
>
> Then bring to mind a person who is struggling and needs compassion. After you inhale a comforting breath for yourself, on the exhale offer the other person feelings of kindness, caring, comfort, and ease.
>
> Then return to yourself, breathing in warm sensations. Then switch back to the person you are visualizing. In an even flow of in- and out-breaths, receive and give warmth and kindness. Like a see-saw. Back and forth, back and forth. One breath in for me, one breath out for you.

You have a compassionate brain that can grow stronger. Understanding the two empathy maps gives you an option to go in one direction or another. As Singer and Klimecki point out, "Compassion is feeling *for* and not feeling *with* the other." This distinction conveys the difference between empathic distress and empathic concern. By recognizing when a moment of suffering arises, taking a pause, and directing loving-awareness, you are increasing the likelihood that kindness will arise in you. So breathe and practice *Presence, Emotional regulation,* and *Perspective taking,* the core ingredients of PEPPIE. Every day offers teachable moments to train your compassion muscle to be more positive, caring, and connected.

Reflection: *With intention and practice, I can nurture and strengthen my ability to be kind and compassionate.*

CHAPTER 12

Face-to-Heart Connections

When the anniversary of the September 11 attacks fell on a Sunday, I was looking for solace and community. So I decided to go to church. It was a blustery and humid day, with rain splattering against the tall windows. It felt cozy and peaceful. *Ahh*, I thought, *the world is quiet here.*

In the middle of the service, the minister's microphone turned off. Alerted to this, she reached for her device and flicked the switch. Boom! The acoustic shudder tore through the church and everyone jumped in unison, hands to ears. We were shocked into silence. It hurt.

Then the lone wail of a little girl filled the hall. Gathering herself, the minister walked off the sanctuary stage and up the aisle toward the crying four-year-old, who was clinging to her father's neck.

"That was a really scary sound, wasn't it?" the minister asked softly, speaking to the child as if she were the only person in the room. "There are other sounds that aren't so scary. We can make them too. Should we try?" The minister began to hum softly to the child. The rest of the parishioners joined in unison. The girl's sobbing faded into song.

Tender Love and Care

That morning, two things stood out for me: the utter stillness needed to recover from the assault of the noise and the compassionate attention offered to the crying child. How nice if we all could be comforted in such distressing moments—by others or by our own ability to self-soothe. Because every single day you, like me, experience noise and distractions from the *outside*. It may not be a piercing boom but rather a steady stream of demands on your attention that's part of living in this modern world: to-do lists, texts, news, social media, job tasks, family needs, bills, rush-hour traffic. These can pile up until you feel bombarded and scattered.

You might also experience distractions *inside* yourself that come as unhelpful beliefs, complaints, rehashes of past disappointments, or wishes for a brighter future because what you have now is lacking somehow. It's so easy to be unkind, berating yourself internally with thoughts like "I have no time for anything that matters," "It's my rotten luck," or even "What's wrong with me?"

Both external and internal assaults can keep you in a constant state of stress that, as I've discussed in previous chapters, can lead to serious harm in your emotional states, health, and relationships. Just as the little girl in the church needed to be tended to after a jolt to her senses, as an adult you do too. Self-soothing is a foundation for emotional self-regulation, and it arises from a very basic need—to be comforted with generous doses of TLC.

Face-to-Heart Interactions

"Compassion for others begins with kindness to ourselves," says the meditation teacher Pema Chödrön. Many wise people insist that in order to love others, we must first love ourselves. Paradoxically, self-kindness doesn't just spring out of nowhere—it arises from being physically held by another person from the moment you were born. Envision the quintessential image of a mother with babe in arms, perhaps a memory of holding your own infant. Physiologically, a bonding process is unfolding within that image—eye gazing, rhythmic breathing, beating hearts, cooing sounds—sealing the human connection. Such tenderness is inherently social and has been part of your neurobiology from the start of your life.

Stephen Porges developed a theory about what he calls our "social nervous system" and our ability to get calm. It's called *polyvagal theory*, which refers to the two *vagus nerves* that begin the brain stem. The *vagus nerve*, the longest nerve in the body with the widest reach, is like a vine that branches throughout your body. It connects the brain with the heart and runs through the throat and neck, and into the gut. It supports the parasympathetic nervous system, as described in chapter 9. One branch of the vagus nerve evolved later and is unique to mammals who need other mammals for survival—yes, humans *need* each other. This newer nerve branch is involved in your body's ability to self-regulate, connect, and communicate with others. It's also involved in your intuitive abilities or "gut instinct."

This nerve branch allows your nervous system to unconsciously detect safety and risk in your environment, which Porges calls *neuroception*. It's your body's way of asking: "Is this person I'm with safe?" "Is it okay to approach him or her?" "Should I be in this situation or should I leave?" The body responds first, the mind later. So when your body automatically responds to stress, it might happen in ways you can't understand, such as fighting, running, or freezing. Sometimes you may feel ashamed about your reaction later: "Why didn't I speak up?" "Why didn't I run?" "I went totally blank." "Why didn't I say hello?" "What's wrong with me?" The stories you tell yourself afterward—fraught with regret, shame, or blame—can hinder understanding and healing.

Awareness of how your body is responding to something is important because it allows you to understand your triggers, and when you have this understanding you won't feel at such odds with yourself. This is the root of self-care. You can even consciously engage the vagal system for soothing. One way is to use your voice. You can speak to anyone, even yourself, in a lulling, affectionate tone, which is called "motherese" or "vocal prosody." That has a profound calming effect physiologically. Like the frightened little girl in the church, you can quickly feel soothed by a gentle tone, whether you're making it aloud or silently in your mind. These kinds of "face-to-heart" interactions are the seeds of emotional self-regulation. So even when you don't have people you love and trust at your side when you need them, you can recruit your own physiology to nurture yourself. All it takes is a bit of gentle self-talk.

People with strong *vagal tone*—a measurement of heart rate variability and stress hormones—tend to be good at calming down and bouncing back from stress, and they report a higher quality of life. And various experiments show that compassionate responses activate the vagal nerve and promote well-being. So be well, tend to yourself, and kindness will flourish.

Kindness in Practice: *Self-Compassion Statements*

To consciously engage your vagal system for self-soothing, start by creating your own compassion statements. Psychologists Christopher Germer and Kristin Neff offer an exercise that can train your brain to be more loving. Similar to the loving-kindness meditation described in chapter 5, you can create messages of kindness to meditate on and repeat, which calms your body and nurtures goodwill toward yourself. The instructions for creating these personal affirmations are simple:

1. Be clear,
2. Be authentic and true to your experience, and
3. Use a kind tone.

Whenever you need bolstering, you can craft a message by asking yourself, "What do I need to feel calm in my body?" or "What do I yearn for from others?" The answers are typically universal human needs: belonging, connection, encouragement, love, patience, protection, respect, tolerance, validation, well-being. For example, while writing this book, creating personal statements helped me assuage self-doubt and anxiety. One statement I wrote was "I have a beautiful message to share with the world. I will speak my truth." When you land on the right phrase, in response you'll sense a twinge of relief, a spark of inspiration, or gratitude: "Ah, this fits for me," "Oh, this feels right," or "Thank you." You can also try one of these on for size and refine it as you go, and switch "I" for "you" depending on how you like to hear your inner voice.

I love myself just as I am.

I will be okay.

I trust in myself.

I hold myself gently.

I've got this.

I am here for me, I am here for you.

I am strong.

I am beginning to feel love and kindness expand.

Even though this feels hard, I will be gentle with myself.

You have the ability to influence how you respond to life's struggles. "Compassion isn't some kind of self-improvement project or ideal that we're trying to live up to," Pema Chödrön writes. "Having compassion starts and ends with having compassion for all those unwanted parts of ourselves, all those imperfections that we don't even want to look at." Drawing upon your inner resources for caring, you can quiet your brain's alarm system, build resilience, and feel a greater sense of empowerment. Porges says, "What we really need to do is inform ourselves about the disruptive things that may have happened to us—not to be angry or to blame but to try to understand the strategies that our body has taken to adapt and to survive. Then we can evaluate whether those are really good strategies." The more you practice kindness for yourself, the more resilient—and compassionate—you become for others.

Reflection: *When I feel upset or uncomfortable, I can calm myself using my body's natural ability to soothe.*

CHAPTER 13

Hugs and High Fives

When I was six years old, relatives in Germany shipped a huge box with two down comforters to me and my sister. The billowing blankets, *bettdecken*, were so exotic and luxurious that we were beside ourselves with wonder. We could sink right into them and be snuggled on all sides. Even today, I sleep best when massaged by the weight and warmth of *bettdecken*, feeling deeply soothed by the simple luxury they offer: touch.

There's so much to the power of touch, which Hollywood captured amazingly in *E.T. the Extra-Terrestrial*. I love watching the unlikely friendship between a boy and a homesick alien develop as they go on adventures to help E.T. return home. In the final scene, they stand in the starlit woods to say good-bye. Quietly, awkwardly, the boy and alien embrace. When they reluctantly pull apart, E.T. touches his own heart, which glows red. "Ouch," E.T. groans. Elliott then takes E.T.'s finger and presses it to his own heart and repeats, "Ouch." When the spaceship doors open for E.T., the alien gently extends his finger to Elliott's forehead and says, "I'll be right here." Cinematic special

effects, like a glowing finger and a red heart, can remind us just how much touch affects our need for love and belonging on a very cellular level.

Our bodies register moments that literally touch us even if our minds do not. These gestures can occur as an ordinary part of any day, with extraordinary effects. As natural as breathing, rarely do we acknowledge what such contact conveys. When my mother-in-law was in the intensive care unit, her best friend happened to also be there, tending to her husband. These girlfriends, Lorraine and Lou, now in their late eighties, unexpectedly found themselves in the same hospital where they had birthed their children. Instead of meeting for an early-bird dinner in town, or one of the Wednesday shopping dates they had been going on for sixty years, they were surrounded by life-support monitors. Lou, deaf in one ear and blind in one eye, leaned in and took Lorraine's small face in her hands, looked her in the eyes, and said, "Lorraine, I love you." And my mother-in-law replied "I love you too, Lou."

My husband's eyes well up every time he recalls the moment of those cupped hands on his mother's face. "I will never forget it. It was brief but so unpretentious, such raw affection between two people. It is profound how deep a relationship can exist between friends for that long." More than anything, that depth was conveyed with a touch.

The Energy of Tenderness

Love and compassion are communicated universally through voice, facial expression, and touch. We rub up against so much in life. The slightest graze of arms or an embrace can convey more than words ever can. It is through touch that we know safety, affiliation, and understanding. The opposite can also be communicated with unwanted touch—those gestures or violations we are likely never to forget or to bury so deeply to avoid the pain of remembering. Yet, gentle touch connects us. As the poet David Whyte describes, touch is an intimate form of meeting.

> Whether we touch only what we see or the mystery of what lies beneath the veil of what we see, we are made for unending meeting and exchange, while having to hold a coherent mind and body, physically or imaginatively, which in turn can be found and touched itself. We are something for the world to run up against and rub up against: through the trials of love, through pain, through happiness, through our simple everyday movement through the world.

As you grew up and faced life's challenges, you may have lost this vital connection to soothing yourself and being soothed by others, especially when you are suffering. This happens so commonly that, when we notice touch, it becomes a rare delight. And yet touch is so primal. Thich Nhat Hanh, the Vietnamese monk, encourages everyone to imagine taking care of themselves as a mother might. When a mother hears the distress of her crying child, her natural impulse is to embrace the baby and—through touch—offer "the energy of tenderness."

Triggering Your Love Hormones

Affectionate touch in childhood profoundly affects how you manage the various ups and downs later in life. Soothing touch, in particular, fosters your ability to regulate emotions. It becomes absorbed into your being, bolstering the nervous system and strengthening attachments.

Tiffany Field has studied a wide range of touch-related topics, from infants to the elderly. Her studies show that gentle stroking has amazing benefits. Even as little as fifteen minutes of tactile stimulation can spur growth and weight gain in premature infants. Across your lifespan, physical massage, however brief, can have a wide range of healing benefits, reducing symptoms of chronic pain, auto-immune diseases, depression, anxiety, and insomnia. It also alleviates mental and emotional symptoms related to addictions, attention deficit/hyperactivity disorder, bulimia, and PTSD.

These effects are so profound, because when you are affectionately and appropriately touched, pressure receptors under the skin are stimulated. This causes a cascade of physiological and biochemical events in the body. Pleasant touch activates the nerves in skin and triggers the reward centers in the brain, including the anterior prefrontal cortex. The activity in your vagus nerve increases and your heart rate slows down, which elicits feelings of relaxation. In turn, this calming effect results in fewer stress hormones and more soothing hormones circulating in your bloodstream—which helps a greater number of immune cells survive. With touch, you get both an immune booster and an emotional pacifier. Any affectionate touch, even self-touch, signals your body's soothing system, channeling positive feelings and a sense of safety.

Ask anyone who has held a baby, clasped hands with a friend, or been embraced by a lover about the rush of pleasure they feel in these physical encounters. It happens because touch spurs the parasympathetic nervous system to release two neurochemicals that calm and connect: oxytocin and vasopressin. *Oxytocin* is produced by the hypothalamus and released by the pituitary gland, and is affectionately referred to as the "love hormone," "cuddle chemical," or "moral molecule." It washes through the body to elicit a sense of euphoria and an intense desire to cherish and protect loved ones. Oxytocin also reduces anxiety and stress, increases romantic attachment, and enables trust and emotional stability. *Vasopressin* supports nurturing and bonding, helps relieve pain, and regulates the body's endorphins—which are a source of positive feelings and can even make us feel high on life.

Touch communicates emotions in a highly nuanced and sophisticated manner that is more accurate than looking at or listening to another person. It conveys at least four negative emotions: anger, fear, sadness, and disgust. And it shares four positive and prosocial emotions: happiness, gratitude, sympathy, and love. Neuroscientist David J. Linden describes skin as our social organ and touch as the social glue. Touch is vital to social experiences including trust, cooperation, compassion, gratitude, kindness, and love.

Touch informs the most common daily interactions even in ways we may not realize. We use it to immediately identify allies or enemies. With appropriate touch, doctors are perceived as more caring, restaurant servers get more tips, and librarians and teachers have students with more engagement and better attitudes. Professional basketball players who engage in fist bumps, high fives, and chest bumps have more personal wins during games and overall better team records. Even rough-and-tumble play is essential for young children in parent bonding and developing friendships. But context matters: how we interpret touch is influenced by our social evaluation of others, including gender, race, or social status. A slap on the back can mean different things depending on the situation.

Despite its importance, touch can be a touchy topic, and cultural trends move us away from healthy doses of it. No-touch policies are common yet run counter to the science on how necessary it is for well-being, emotional regulation, and bonding. The risk is that the more we are conditioned to avoid touch, the more we split our minds from our bodies and our bodies from others' bodies. When we lose touch, we lose our sense of connection. Allowing ourselves to be touchable and to touch others in mutually respectful and affectionate ways grounds us in the present moment, stirring the energy of tenderness.

Kindness in Practice: *Getting in Touch*

Fifteen minutes a day of touch is restorative, and you can give it to yourself. Practice self-care while savoring sensations like the warmth of tea, the cascade of water during a hot shower, the soft fur of a pet, the smoothness of a polished stone, the cuddliness of fleece. At any time, there is a way to evoke the powers of touch. Just get creative. When you are struggling or feel tense, try giving yourself a gentle squeeze. Touch your face. Place your hand over your heart and "touch in" with your feelings.

It helps to spend a few days noticing how often you touch others and are touched by others, be it a handshake, pat on the back, embrace, or kiss. Notice the differences you experience from day to day. Some days may be low-touch days and others high-touch. Just observe for now—and notice your responses. Over time you may clue in to your optimal touch barometer:

- Do you find that you like to be touched or not so much? Consider the culture of your family or community when you reflect on this type of affectionate expression.

- Do you see any changes in your mood, energy level, and quality of relationships that correlate with the amount of touch you've received?

- Do you desire touch in some situations but avoid it in others?

- Do you notice things that touch you emotionally? If so, what thoughts or feelings are evoked?

Naturalist Diane Ackerman writes that "Touch seems to be as essential as sunlight." Touch encourages you to be aware of sensations and is another way to bring kindness into your direct experience. You won't always have connections with others in moments of need or desire, but you can be tender with yourself. The experience of touch is more than skin deep: you emotionally tuck yourself in to comfort and use it to purposely recode your caring circuitry. By reaching out to others, allowing yourself to touch and be touched just a little more often, you will contribute to the conduit through which kindness can flow. So touch with kind affection, empathy, and sensitivity—and break the SPEL. In doing so, you discover another way that your body is a sensitive instrument capable of responding skillfully and kindly to your inner and outer worlds.

Reflection: *Allowing myself to be touched and to touch others sparks the energy of tenderness and connects me to humanity.*

CHAPTER 14

Taking in Kindness

I met Joey when he was twenty years old. He had an unusual history: as an "out of control" child, he was among the first children under five to be treated with medication for bipolar disorder. "I was a little guinea pig," he recalls. Joey also carried the psychiatric labels "oppositional defiant disorder," "conduct disorder," and "severe attention deficit disorder." School had been difficult, and his self-esteem was fragile.

When he started seeing me for therapy, I saw an earnest young man trying to make his way in the world with good intentions. Joey would show up to sessions in cut-off T-shirts and gym shorts, proud of his sculpted physique. He reported that he was entering body-building competitions. I was curious and asked him about it. This is what he shared.

As a teenager, Joey weighed three hundred pounds and had been the fattest kid in school. He attributed his weight gain to side effects from medications and indulgent parents who loved him with generous portions. He struggled with his weight until something extraordinary happened at school. A student he didn't know well offered to meet Joey on the track before school to run laps together. It wasn't another cruel joke; the other boy meant it. And Joey took him up on the offer.

"I could barely walk around the track," Joey says. "But every day I met this kid, and he'd walk with me until eventually I could run a bit." The new friend kept showing up. Joey kept at it. For the first time in his life,

Joey understood that he had control over his body, and that when he put his mind to something he could do it. All because of one compassionate kid.

"Running laps" became the foundational metaphor for other challenges in Joey's life. If he could do it once, he could do it again. It took patience. He also came to understand the power of compassion his track buddy had modeled. Over time, Joey discovered a desire to work with "out of control" kids like he had been. He took classes at a local college and started working in schools as a behavior specialist. No doubt, Joey could run with it.

Choosing Positive Change

Gestures of kindness can come when we least expect them, from the unlikeliest of sources. Sometimes the hardest things are to accept help and to trust that the help will benefit you. Deep inside you, there's a wise, loving self who knows—as Joey's friend did—what is good for you. And the kindest thing could be the hardest to do. This wise self is the voice of courage that can get you through any "should," "ought to," and "no you can't" judgments and doubts.

When you step through the vulnerability of needing help and walk past the fear of being exposed, your heart cracks open to new possibilities. The novelist James Baldwin wrote, "Not everything that is faced can be changed, but nothing can be changed until it is faced." Sometimes you need another person to believe in you, to stand face-forward with you and be the voice that says, "I'm with you. You can do it." Until one day, after sweating it out over and over, you believe, "Yes, I really can."

Positive Conditioning

Joey's story shows what can happen when you choose to make a change and stick with it, with love and acceptance. This is the power of *positive neuroplasticity training*, what psychologist Rick Hanson refers to as "taking in the good." New connections in your brain are being formed in every moment and through every interaction you have. The more you expose yourself to negative or harmful influences or habits, the more they stick. In the same way, the more you expose yourself to positive things and habits, the more beneficial experiences stick and become lasting inner resources. In other words, what you choose to focus on and practice will grow stronger. This means you can influence your own brain on a very deep, cellular level *on purpose*.

Taking in the good cultivates more good in your life. Focusing on experiences that encourage well-being offsets the negativity bias inherent in the base model of your brain. So if you want more kindness, compassion, love, gratitude, joy, forgiveness, and trust in your life, then you must intentionally grow those states of being. How? By engaging and training in experiences that promote kindness, compassion, love, gratitude, joy, forgiveness, and trust! Not just once or twice—but *repeatedly*. Run those kindness laps and strengthen caring neural pathways.

Hanson teaches that there are three ways to take in the good as part of everything you do, whether daily interactions with family and coworkers, spending time in nature, or traveling to explore other cultures. Here's how to do it.

- **Notice or create a beneficial experience.** Beneficial isn't synonymous with pleasant. Joey didn't exactly enjoy running laps, but he knew it was good for him and that the progress felt good. Other beneficial experiences include going to bed on time, reading a good book, playing with children, meeting a friend for coffee, or listening to relaxing playlists during a commute.

- **Be present.** Stay with the experience by noticing the sensations and imagery. Don't let the good moments pass you by.
- **Let the experience stick in your mind.** Savor it. Intentionally recall it, and, when you do, experience the positive feelings all over again.

Kindness in Practice: *Your Kindness Plan*

What is one thing you can do every day to slowly shift the neural patterning in your brain toward positivity and self-care? Write this out in your journal, keeping it short so you can read it quickly. Then tape the intention to your bedroom wall, bathroom mirror, refrigerator, or office space so you can read it regularly, as repetition is important. Here's a kick-start for how you might express your desire for what you want to change and how you want to go about getting it. You can copy the script below and fill in the blanks in your journal.

> I want more [friends] so that I feel [connected]. To be the person I want to be, I'd like to feel [the empowerment that comes with support and comfort]. One positive step I will take today is [to talk with one person at the coffee shop]. I promise to schedule time to do this step [every morning when I buy my coffee]. I may even let [my brother] know that I'm taking this step so he can support my efforts and encourage me on days when I may fall short. I know that anytime I choose to take this positive step of [talking to a new person], I am building up fresh neural pathways in my brain and growing new inner strengths. I promise to be [kind and caring] toward myself every step of the way.

Growing a kind mind takes practice, and practice makes progress. You can direct inner change, and that is empowering. Joey was laying down new neural connections every morning at the track, and what he learned helped him years later. This ability is an invitation to take gentle care of your brain by taking in the good—consistently—to sculpt your experience. Simply remembering that any positive or negative sensations, thoughts, and feelings that you focus on will shape the inner landscape of your brain is empowering. Knowing this allows you to choose kindness over criticism, faith over fear, and understanding over indifference.

Reflection: *Because I have the power to bring positive change into my experience of myself, others, and the world, what do I want to cultivate?*

PART 3

Kindfulness

noun kind · ful · ness \ ˈkīn(d)-ful-nəs \

: being aware of the present moment with heart

CHAPTER 15

Mindfulness with Heart

Life can get so busy that it flies by. Sometimes it's only later when we realize we missed something along the way. I learned how my daughter Josie earned her childhood nickname, Bugsy, only recently, when we took our beloved former babysitter out for dinner to celebrate her long-awaited green card. Pauline took remarkable care of our children, with a flowing sense of time, an ability to find their antics utterly hilarious, and a smile so large that no one in her presence can take life too seriously.

At dinner, Pauline shared a story that took place fifteen years earlier: "One day I was doing the laundry. I didn't hear Josie, and I wondered, *What is she doing?* I went into the playroom and—oh my goodness!—there were ladybugs everywhere. Little red dots. And she was eatin' them. I said, 'Josie, no! Don't eat the bugs.' It was weirdly cute though, because she was just sitting there picking them up one by one and eating them. Like a snack! She was just in her own little world and happy. Then I said, 'Okay, your name is Bugsy. That is going to be your name from now on. Bugsy.'"

Busy with the hectic pace of life in those days, I'd never paused to ask why suddenly everyone was calling my daughter Bugsy. It just seemed sweet, so I went with it. Now, listening to Pauline, clearly a cherished *allomom* to my girls, I realized how I had taken for granted the gifts right in front of me. Because even the small things—like ladybugs, like cute little nicknames—can be so important. When we are not mindful, we can lose these quirks and joys in a sea of distractions.

Being Present to Your Life

The practice of mindfulness is not really about having a healthier brain or body, aging gracefully, or focusing at work—all ways our consumerist culture brands it. The true motivation for mindfulness is simply *not to miss your life*. Jon Kabat-Zinn, a biologist and practitioner of Zen Buddhism credited with bringing mindfulness to mainstream medicine, refers to the experience of modern life as "full catastrophe living." Our constant state of information overload, distraction, and stress leaves little room for the time and stillness we need to quiet the mind and body, appreciate everyday pleasures, and ignite our caring circuitry. No wonder we're under a SPEL, caught in *Self-Protective Empathy Lethargy*, trapped by just coping.

Mindfulness can wake us up from that. It is the awareness that emerges through paying attention as each moment unfolds, with intention and without judgment. We evoke it by noticing what's in our field of awareness—without getting attached to the feelings, thoughts, or images that appear in our mind's eye.

Mindfulness is at the core of many ancient wisdom practices including prayer, meditation, chanting, yoga, tai chi, and qigong. Teaching it to patients of Western medicine had revolutionary effects on those patients' well-being. It is now a mainstay of mind-body approaches to health and a widely accepted practice. What is less widely known is that *mindfulness* is essentially a practice of *compassion*, for both yourself and others. In the Chinese ideogram, "mindfulness" consists of two symbols: one represents "presence" or "now," and the other represents "heart" or "mind." So mindfulness can also be translated as "presence of heart" or "heartfulness." Other expressions—"caring mindfulness," "affectionate attention," or "loving-awareness"—also highlight the connection between compassion and presence.

My favored term is "kindfulness," which I define as *experiencing the present moment with heart.* We can take ourselves much too seriously in mindfulness meditation and forget that the energy of kindness is at the heart of the practice.

Part 3 explores different ways of seeing and doing kindfulness—whether through nature, music, touch, prayer, play, or developing an eye for each day's tiny moments of connection, presence, and delight—so that we notice things like ladybugs. Kindfulness brings a loving spirit to mindfulness as you look at your experiences through the lens of curiosity, wonder, and appreciation. You can shift from stressed to blessed.

Kindfulness Loves Company

The popularity of mindfulness is due to one important fact: mindfulness practice works. The science, from hundreds of studies, is clear. Mindfulness skills, such as the one offered below or the "loving-kindness meditations" offered in chapter 5, help with a host of life challenges, from anxiety to chronic pain. Neuroimaging scans show brain activity changes, as mindful attention exercises improve emotional regulation, attention, and responses to painful experiences. Training in mindfulness may lead to enduring changes in the brain—even among people who have just learned basic skills in eight weeks with as little as twenty to thirty minutes of practice a day.

It's worth noting that most of the research on mindfulness cited above occurred in settings where people practiced, at least some of the time, in groups. For people new to mindfulness meditation, it's easy to presume that the techniques are too simple to have any real benefit. *Three minutes of breathing? Visualizing a calm lake? Walking slowly? Really?* Or they can feel too hard to maintain. *My mind wanders. I can't sit still. I fall asleep! I get bored. I'm bad at this.* Unsettling inner dialogues and unpleasant feelings become more apparent, which can be dismaying. Learning the techniques in a group setting means learning that *other people* are struggling with chronic pain, heart disease, cancer, anxiety, depression, loss, or stress—just like you. You come to know that you are not alone in your pain or distress, and you see that others also find a mindfulness practice to be challenging at times. Mindfulness groups or classes create cultures of kindness and safety. Social support can also make things stick—a kind of "we're in this together" mentality as you flex your compassion muscle. If you encounter a resistance to the training, a little help from a kindfulness buddy can go a long way. These days you can find strength and accountability in classes and by using meditation tracking and support apps.

Kindness in Practice: *Easing into the Present Moment*

Many kindfulness practices can be done anytime, anyplace, without need for a cushion, pew, or yoga mat. These are intentional moments when you pause to witness your own mind and body with caring attention and compassion. Using the metaphor of a camera with a zoom lens, *zooming in* on the breath can spark the body's natural soothing pathways and ignite a relaxation response. *Zooming out* to more expansively observe your thoughts, feelings, and bodily sensations without judging them makes you a compassionate witness to your experience. Here is a simple kindfulness practice to try:

> Begin by finding a comfortable seat, standing quietly, or lying down. With a sense of grace and strength, start to feel the support of the earth or floor under you. Bring kindful attention to your breath, using it as an anchor, or bring your attention to your feet or your body as you inhale and exhale. Begin to notice a sense of support and ease. You may voice an intention such as "My heart is open to the whole of my experience."

> Then start to open to your field of awareness, noticing your sensations, feelings, and thoughts, how they flow before you and through you. Like a passenger on a train, begin to notice the expanse around you, swaths of experience: colors, textures, sounds, sensations, words, images.
>
> As you do this, you may encounter fluctuations or waves of sensation. It may be quiet, still, or busy, flowing back and forth from stillness to commotion. Whatever you notice, it is just fine.
>
> If you notice pleasant sensations, feelings, or memories, simply acknowledge them. If you notice unpleasant feelings, sensations, or memories, acknowledge them too. Experience these moments with tender and kind attention.
>
> Whatever arises remains anchored by your breath, or your feet, or the ground beneath you. Stay centered in the vastness of the experience. Notice how kindfulness allows you to receive, without judgment, and to be gently present to the whole of your life right now. When you are ready, turn your awareness back to your surroundings, feeling that deep in your center, all is well.

With practice, you become able to view your stream of thoughts, emotional states, and bodily sensations as a participant-observer—as if you were in a movie theater watching a screen and yet also in the scene. The meditation teacher Jack Kornfield writes, "In the present moment we can learn to see clearly and kindly. With the great power of this mindfulness, we can become fully present to the unbearable beauty and the inevitable tragedy that makes up every human life." This happens because, over time, these simple practices reveal that you are essentially okay and that deep down, under the layers of self-critique and life's hard knocks, you are perfect and whole. Kindfulness removes the veil of separation between you and others so that you can hold your own humanity with loving-awareness.

"Who you *think* you are and who you *really* are, are infinitesimally close and also infinitely far apart. And there is enormous suffering in the distance between," says Jon Kabat-Zinn. "When we get in touch with what is deepest and best and beautiful in ourselves *already*, then we actually don't have to get anywhere else. It's the beautiful mystery of our lives." The heart of these practices shows us that we can face everything in life with a kind heart.

Kindfulness is a communion of the qualities of mindfulness and kindness, awareness and love. It helps you to connect your head with your heart so that you don't miss out on your life. You can transform fear into faith. When you approach life with kindfulness, you are fully present to everything: gritty and graceful, painful and joyful, effortful and effortless.

When still under a SPEL, you may think that all this mindfulness seems charming enough, but life is hard, people are mean and selfish, and golden rules have gone out of fashion. There are bullies in pulpits and parliaments, kids strapping on suicide vests, refugees drowning, and people losing jobs. By becoming present through mindfulness, you can see that there are also teachers in trenches, children leading causes, and volunteers building homes. There are ladybugs you can notice with delight. There is incomprehensible devastation and there is untold beauty. It's all there. You can break the SPEL by being present for the full range of experience and having kindfulness for your straying, distracted, beautiful self and for the similarly challenged people in your life.

Reflection: *It's up to me to approach every day with kindfulness and to notice how wonderful life can feel when I awake to it.*

CHAPTER 16

Your Loving Self

It's true that you are a wise self with a core essence of love and kindness. But this basic fact can seem far removed from your awareness. It may even seem ludicrous because it is buried under so many difficult emotions and past pains. As Dina pointed out to me, you can't *think* your way to this core nature, you must *experience* it.

When Dina found her way to a recovery program, her life was in pieces. "Nobody knew that I was waking up next to guys and I didn't even know their names. I was stealing money and lying to everybody. If someone would have said, 'You're not accessing your inner source of power, peace, and wisdom,' I would have told them exactly where to put that idea. I was suicidal. I was alcoholic. I didn't see the essence of anything. I didn't see hope in anything."

Then the woman Dina chose as her coach gave Dina one simple instruction before starting the first step in her recovery: learn to meditate. Dina rolled her eyes in response. But her guide didn't even blink, she just said, "Your way doesn't seem to be working. Why don't you try mine?" The instruction was to sit still every morning for twenty minutes and focus on her breath. Dina spent the first few attempts curled in the fetal position. "I would just shake and sweat. I could not sit still for more than three minutes at a time. So I considered how many three-minute periods I would have to do in order to make the twenty-minute quota. And I did six, seven of these short meditations throughout the day. I thought my coach would be mad at me, but she said, 'Whatever you can do right now is fine. Just don't drink.'"

At first, Dina didn't think her brief sessions did anything, and yet she didn't mind them either. A tenderness toward herself was emerging: "I had a feeling for myself that you feel for a child. Even with all the bad stuff I had done, I felt like saying, 'It's okay, Sweetheart. It's okay. You are safe.'" After eight weeks, Dina had an experience of awakening when she saw there was more to herself than she could possibly grasp.

Dina found her way to healing through small, mindful moments. Collectively, they led her to forgive her past, trust her own wisdom, and nourish her spirit. "Honestly, the only thing I ever did was three-minute meditations. Everything was born out of that." Her memoir, *Madly Chasing Peace,* describes how she renews compassion for herself all day long. "Being able to clear the past just keeps the channel open—the channel to that inner wisdom, to that intuitive nudge, *the whisper*." In this way, Dina's heart grew strong in kindfulness. So can yours.

Meeting Your Suffering with Kindness

Dark nights of the soul are times when we come face-to-face with our past pains, and they can be incredibly hard to endure. The only way to heal through them is being still, present, and kind to the difficult feelings. This creates an opening so that self-compassion arises. Jack Kornfield writes, "When we stop fighting against our difficulties and find the strength to meet our demons and difficulties head on, we often find we emerge stronger and more humble and grounded than we were before. To survive our difficulties is to become initiated into the fraternity of wisdom."

Kindfulness allows for *integration* of unwanted aspects of yourself. The neuropsychiatrist Dan Siegel sees integration as linking the mind and the heart into a coherent whole—and as the source of healing, harmony, and well-being. He writes, "For the mind, integration means kindness and compassion." Dina, who has now been meditating for more than eight years, agrees. "All I was saying in my meditations was: 'I'm open. I'm willing. Just show me. I surrender doing it my way. If there's another way, I'm open to seeing it.' Self-compassion was born through being in that state."

Self-Compassion Stirs Old Pains

We know light because we know darkness, we know hot because we know cold, and we know pleasure because we know pain. Christopher Germer and Kristin Neff, cofounders of the Mindful Self-Compassion (MSC) skills training program, point out that the mind needs contrast to know anything. So when we begin to gently look within, develop warm wishes toward ourselves, or use tender phrases such as "May I feel safe" or "May I be at peace," we naturally bring to mind past experiences of pain or times when we were not treated kindly. This can activate the sensations of suffering and feelings of self-hatred or shame. Germer uses a fire's back-draft as a metaphor for how this happens. When a fire is deprived of oxygen and a burst of fresh air blows in, the fire expands—often with a roar. If old pains are burning under the surface of our consciousness, tender feelings toward ourselves are like the explosion of air. The old pain roars forth—that's back draft. As Germer points out, "When you give yourself unconditional love, you discover the conditions under which you were unloved."

The vital thing to keep in mind is that, as the old pain becomes exposed, it is presenting you with the opportunity to heal through kindfulness. Germer adds, "When we say, 'this is a moment of suffering,' we're actually creating perspective. That's a mindful moment. This is a moment to actually choose to be kind." This not only heals past pains, it makes us happier and more able to be kind to others.

In the MSC program, participants meet for two hours a week for eight weeks. They learn the basics of mindfulness and self-compassion practices (such as "loving-kindness meditation" described in chapter 5, "hands on heart" in chapter 11, and "self-compassion statements" in chapter 12). When the effects of the MSC program were studied, comparing people who completed the program with those on a waitlist, the results were remarkable. By building up inner resources to face their suffering and the suffering of others, people in the program became more compassionate to themselves and others. Moreover, there were sustained psychological benefits over time: increases in self-compassion, mindfulness, and compassion for others; greater life satisfaction; and decreases in anxiety, depression, stress, and avoidance.

Self-compassion skills training has been shown to be beneficial for a wide variety of people across lifespan, ethnicities, and circumstances, including teenagers, war veterans, and caregivers. All these benefits come from choosing to be kind to our suffering selves. And this nurturing skill takes just minutes a day to develop.

Kindness in Practice: *Small Doses of Self-Kindness*

Like Dina discovered, a repetition of small doses of meditation practice can be a powerful component in enabling you to heal. This is a wise approach, as it reduces the chance that you will give up on a more formal meditation practice that takes twenty minutes or longer. Rather than getting overambitious, feeling "I can't do mindfulness," experiencing a sense of failure, or giving in to long periods of daydreaming, keep it simple. Shorter repetitions throughout the day encourage the positive neuroplasticity in your brain to grow new connections. As Germer notes, "Since most of our suffering occurs in daily life, cultivating informal three-minute practices throughout the day *potentiates* self-compassion." Graciously, Dina offers you this meditation:

A Gentle Embrace

Let yourself settle into a place of comfort and relaxation.

Allow your mind to come to a place of stillness and of peace.

Let the muscles in your body melt, melt, melt into absolute relaxation.

In this moment, let yourself find a moment of freedom from anything that's been weighing on you lately.

Feel your breath gradually becoming deeper and stronger in your chest and in your belly.

Anchor in the knowing that when you connect to your breath, to the energy and love that are always there in your heart, that your intuitive guidance, the power greater than yourself, is always there as a place of comfort, freedom, and wisdom.

Your highest self, the highest intuitive knowing and energy within, is always big enough to absorb anything you'd like to let go of.

Settle into that gentle knowing and healing space that your heart energy emanates.

Let yourself breathe in deeply, feeling the weight of your burdens melt away and allowing a kind and comforting presence to embrace you in absolute peace, serenity, and freedom.

When you're in an environment of love and compassion, you get a chance to heal. In a way, you get a chance to re-parent yourself. To be a kindness warrior. To put love in action for yourself. It's very hard to fall under a SPEL when you cultivate self-compassion because it boosts empathy and empowerment. The skill of self-compassion is like having a kind, caring protector around you all the time who will help to meet your pain and suffering, shame and vulnerability, and transform it into something beautiful. It allows for integration of the unacceptable parts of the self with new experiences that are intentionally nurturing and beneficial. This builds resilience and opens the channel for gratitude, forgiveness, love, and purpose.

For Dina, "It's finding that depth of center, that inner source of power, peace, and wisdom. To know that that deep essence is within everyone. Sometimes it is hard to notice—it's like a whisper at a rock concert." That whisper can help you kindly accept who you are, and who others are, unconditionally.

Reflection: *With self-kindness, I can honor all the tender parts of me, the ugly and the beautiful, and come to feel how lovable I am.*

CHAPTER 17

The Naturalness of Being

I have long held the image of a garden as a metaphor for growing a kind mind. It's not a new metaphor, as sages, poets, mystics, and scientists have also been drawn to relate nature to our inner, subjective life. Nature was a potent healing image for Ivy, a young woman who had a particularly hard time staying present and practicing kindfulness.

Ivy developed a coping strategy that once served her well in the face of childhood abuse: she would dissociate from her body. (That's the freeze response to traumatic stress mentioned in chapter 8.) In the safety of a support group, she experimented with awareness and grounding practices, but it was never easy for her. Then she shared something with the group that was helping her heal. It was an animated short film called *The Man Who Planted Trees*. This fictional story is about Frenchman Elzéard Bouffier, who, after the devastation of World War I, single-handedly grows a magnificent forest from a pocketful of acorns. As the trees spread over many years, the animals arrive, the waters flow, and people come to the village—bringing the land back to life. It is the ultimate metaphor for the power of nurturing, happening one person, one moment, one day, one year at a time. For Ivy, recovery progressed one acorn at a time.

By befriending nature, we offer ourselves a great gift: a sense of connection with all living things. Children do this intuitively, as I am constantly reminded. Early one summer morning, I was surprised to see Josie sitting as still as a garden gnome by the bird feeder. Oddly, she was wearing winter boots that were sprinkled with birdseed. Sure enough, a chipmunk scurried over her toes and a house wren paused on her knee. She sat like a rock for a good twenty minutes. When she finally came in, I asked her what she had been doing.

"I was waiting for Charles," she said, "the woodchuck."

Josie was completely in tune that day, and she looked content on such a deep, inner level that I truly appreciated the curative power of being in nature. Ivy, too, knew deep down that her own healing was a cycle of life that needed tending. She just needed to give herself the time.

Nature Completes Us

In this day and age, when so many of us live in unnatural spaces and are tethered to technology, we can easily lose connection to our environment and to each other. The lessons from nature can be lost on us. But nature can be a great teacher if we pay attention, as German forester Peter Wohlleben tells us. He wrote a charming book called the *Hidden Life of Trees.* Wohlleben observed that trees, like humans, develop friendships and communicate in caring ways. "A tree can only be as strong as the forest that surrounds it," he writes. Some tree families are so intertwined at the roots that they can die together. Loner trees suffer in isolation, trees use scents and electrical signals to warn other trees of potential attack, and there are jungle rules among trees that promote overall forest well-being and reciprocity. Sounds wonderfully similar to humans. In nature, there is a communal life consisting of all living things that expresses connection and inspires reverence.

Our Bodies' Natural Restorative

The simplest way to expand kindfulness is to tune in to pleasant sensations on purpose by being in nature. Scientists continually find that expansive experiences in nature evoke positive emotions—such as awe, beauty, and wonder—that promote health. Some scientists even argue that nature's curative effects are deeply woven into our human fabric, right down to our DNA. After all, human beings have been roaming the planet for ages, and studies show that we unconsciously respond to nature in ways that calm our bodies and minds. Nature engages your body's parasympathetic nervous system, sometimes called the "rest and digest" system for good reason: nature helps to soothe you. And it doesn't take much. A twenty-minute walk in a park triggers positive changes in heart rate and mood.

Being in nature helps you get out of your head by decreasing rumination, which is the negative, self-critical, and pessimistic thoughts that are a core part of depression and anxiety. In one study, people's brains were scanned with fMRI machines before and after being outdoors for ninety minutes. They reported less rumination and showed less activation in the part of the brain associated with isolation.

Even listening to nature sounds after a bout of stress restores the body to balance, and being in a hospital room with a view of trees or even pictures of nature speeds recovery from surgery and reduces the need for pain medication.

Nature also hones attention, inspires creativity, evokes awe, and spurs kindness. When you aren't multitasking, your brain has a "default network" in the prefrontal cortex that has the space to engage when out in nature.

It's the part of your brain that is imaginative, creative, reflective, and makes meaning of life experiences. Giving your brain a rest by being in nature can ignite new ideas and solve problems.

Scientists at U.C. Berkeley carried out a series of studies demonstrating that the psychological benefits of exposure to beautiful nature aren't limited just to feeling good but also to doing good. For instance, in one of the studies, people instructed to play video games were challenged to give up or share points; those exposed to beautiful scenery were more likely to share than those who saw scenes with less natural beauty. Even being in a room with plants can lead to greater helping behavior, as demonstrated in another of the studies, which told participants that they could support an earthquake relief effort in Japan by folding paper cranes. Generosity was measured by how many paper cranes each person made; the people who were exposed to a room with more greenery felt happier and created more paper cranes.

You can deliberately connect with the natural world as a far-reaching gesture of kindness: to appreciate and care for it while allowing it to sustain you. When you feel better, you are more likely to express kindfulness to others.

Kindness in Practice: *Connecting to Natural Beauty*

Nature is restorative and opens us up to a deep sense of connection. Be kind to yourself by evoking the calming effect of nature in daily life. Wherever you live now, whether you are a city or country dweller, consider where you might connect with nature. Make it a goal to be in nature and among natural scenery as often as you can. Here are some ideas for doing this:

- Set up a bird feeder and watch the birds, chipmunks, and squirrels gather.
- Create an aquarium full of vibrantly colored fish and coral.

- Hang images of natural scenes on your walls.
- Play a nature soundtrack.
- Run a relaxation water fountain.
- Gaze at the stars or the moon.
- Walk through a park noticing trees, birdsong, insects, sunlight, and breezes.
- Visit NASA's online image galleries to view galaxies, or join cloud spotters at the Cloud Appreciation Society to view cloud formations seen around the world.
- Create a top-ten list of films or websites that inspire awe for the power of nature and wildlife.

You don't need to go anywhere to conjure up the beauty of the natural world. Here is a visualization to try anytime, anyplace:

> If the mind is like a garden, then it can be helpful to imagine tending to it with gentle care. Picture yourself in a natural sanctuary of your own making, such as a secret garden. In your mind's eye, create a sacred place to dwell in, a loving and welcoming place surrounded by nature. Maybe you envision yourself in a cozy home garden—a safe place on a strong foundation. Perhaps you are in a lush jungle, a treehouse, or a greenhouse filled with orchids. No matter how messy or sparse your garden seems, simply settle into the beauty of it. It's buzzing with sensations. What are they? Notice the colors and fragrances. Now breathe into this sacred space with a soft and natural breath over and over again for a few minutes. When you feel at peace, bring your awareness back to your surroundings, knowing you can always come back to rest in your own inner landscape.

Watching my daughter's stillness as she waited for animals to approach her was a reminder of how nature puts us in touch with our innate kindness. Rainer Maria Rilke wrote: "If we surrendered to earth's intelligence we could rise up rooted, like trees." Like trees, you were born with an inner resilience that thrives in communion with nature and one another. In nature, you can expand awareness, tune in to pleasant sensations, and feel connected. These are ways of calming the nervous system that will allow you to become comfortable with uncomfortable sensations. Your body is always trying to tell you something important about yourself and the world around you. It deserves a lot of respect. By focusing on your body, which you'll do for the next few chapters, you can cultivate kindfulness from the inside out.

Reflection: *Nature does not judge me. It's a kind and patient witness to my unfolding.*

CHAPTER 18

Radical Acceptance

We tend to be conditioned to conform to narrow views of who we should be. We're also taught to judge one another. This is why unconditional acceptance can feel radical. Elyse learned the power of acceptance from her oldest child. "One dream at a time, I began to realize that my child did not get the script that I had written for her life," writes Elyse. "With each symptom that manifested and took over my child—anxiety, depression, stuttering, stammering, hair pulling, skin picking, sleep disorders—each part of my dream was being chipped away."

As Elyse shared her story with me, a mother of two girls, I could only imagine the yearning and the confusion. Elyse had expected princess birthday parties, mani-pedis, and mother-daughter dates—no such luck. When it came time for the father-daughter dance in elementary school, just finding something for her daughter to wear was traumatic. When she got her period, it was a nightmare. Then there were hospitalizations and times when the household was so stressed that Elyse emotionally checked out. After all, she had two younger children who needed her. One day, her fifteen-year-old daughter texted her mom with vital news: *I am a boy.*

After the declaration, as the pain of the past ten years instantly came into focus, Elyse paused for a long time. "Then I said to my kid, 'Let's get your hair cut.' When the hair he never combed anyways was chopped, it was like a light went on in him. I could see the life coming back, so I thought: *Okay. We can do this.*"

The family was in unchartered territory—Caitlyn Jenner and a nationwide LGBTQ bathroom bill discussion were still to come. So they rallied to learn as much as they could, consulting with experts and doctors. Her son, Landon, was the first trans kid to come out in their neck of the woods—a small, sports-minded, socially conservative community. The school had never been through this situation before, and other parents didn't know how to explain things to their children when one day a classmate showed up with a new name and gender.

It was people's pity that really got to Elyse. She told me, "I had just learned about my kid. I was having lunch with a friend; our kids grew up together. She dropped her eyes and shook her head, saying, 'I feel so bad for you.' I know she meant well because I know her heart, but it bothered me so much I can't even express it. My response was, 'You know what? This is just our path. It doesn't mean there's something wrong with us. It's just a different walk we're taking.'"

Elyse recognizes the challenges others faced, as judgment and prejudice run deep. Yet she chose to stand strong in unwavering love for her child. "I never push my agenda on people," she said to me. "I'm not trying to convince anyone to accept my kid, but if you look at us as human beings, I don't see why you wouldn't. We all really want the same things at the end of the day. We want our kids to be happy and healthy and productive members of society. That's not different for me. I know in my heart if I didn't accept Landon, my kid would be dead. I don't know of any parent who would choose that over clinging to beliefs. Common ground is there if people want to see it."

Listening to Understand

Acceptance begins with the understanding that comes through listening, as Elyse showed when she fully took in the fateful text message. "We think we listen, but very rarely do we listen with real understanding, true empathy," said Carl Rogers, the humanistic psychologist. "Yet listening, of this very special kind, is one of the most potent forces for change that I know." Rogers believed that empathy is not a state of being but a process of deep sensing and listening. And since it is a process, it is something we can practice and cultivate.

Fortunately, life gives us many opportunities to practice what Rogers described as "entering the private perceptual world of the other and becoming thoroughly at home in it. It involves being sensitive, moment to moment, to the changing felt meanings that flow in this other person, to the fear or rage or tenderness or confusion or whatever that he or she is experiencing." With this kind of empathy we come to see as another sees, and everything he or she does makes sense. Just like Elyse, we can reach complete acceptance of another person. As Rogers says, "To be with another in this way means that for a time being you lay aside the view and values you hold for yourself in order to enter another's world without prejudice." Allowing someone to be as they are a great act of kindness.

Compassionate Communication

Communicating in ways that foster understanding and acceptance goes far beyond ordinary, everyday communication. It requires a radical transformation of relating. More than forty years ago, Marshall Rosenberg developed a method called *nonviolent communication* (NVC), sometimes referred to as compassionate communication, which has been used in education, psychotherapy, social justice, and conflict resolution situations all over the world.

In our culture, we are conditioned to understand each other in dualistic categories that are static, confining, and deadening because they are moralistic judgments: judgment/reward, good/bad, right/wrong, smart/dumb, normal/abnormal, winner/loser. Rosenberg called this "life-alienating communication" and believed it to be the source of violence.

When fraught with defensiveness, anger, shame, avoidance, or retaliation, communication can kill kindness.

With an emphasis on deep listening and a quality of presence rather than just spoken words, NVC is a radical approach because it allows for a mutual flow of communication. Rather than focusing on personal evaluations, NVC focuses attention on human needs. Because when needs are not met, then anger, depression, guilt, and shame show up. When needs are met, the result is more pleasant, playful, and joyful. Our feelings result from interpreting an experience based on underlying needs: to be accepted, loved, heard, seen, appreciated, touched, safe, independent, creative, inspired. Sometimes we don't know what we need because no one ever asked us and we don't ask each other. We are not taught to be clear about our needs nor how to express our feelings. Once you're aware of underlying needs, then you begin to take responsibility for the feelings, and you care that another person's needs are met. This requires loving patience to step back, to watch and learn, and to restrain from rushing to judgment.

Kindness in Practice: *Connecting with Kindness*

This reflection exercise invites you to consider the ways you have been conditioned to analyze and judge yourself or others. It's also meant to encourage you to explore how you might practice a new way of relating.

Call to mind an encounter in which either you or the other person became upset. It could be a quibble about whose turn it is to take out the trash, a complaint about time spent texting, someone feeling unappreciated, or a perceived criticism. Follow along with the four basic components of NVC as you reflect on your interaction. As an example, here's how I imagine Elyse might have completed this exercise, considering her exchange with her well-meaning friend, Mary.

Observation. An observation is a clear description of the event. The challenge is to notice the situation without judging it, because if we mix evaluation with observation, it can result in criticism. When we simply observe what happened, the situation remains open.

Evaluation: Mary had the nerve to feel sorry for me.

Observation: Mary said she felt sorry for me.

Identifying and expressing feelings. This means having a diverse vocabulary of feelings, as encouraged in chapter 3, and understanding the difference between feeling and thinking. By taking responsibility for our own feelings when receiving criticism or a negative message, we avoid dropping into the blame game—either of our self or the other person.

Blaming expression: I feel misunderstood. I feel you don't get me at all.

Feeling expression: I'm sad to hear you feel sorry for me, because I was hoping to share how relieved I am that my son is now thriving.

Acknowledging the needs at the root of feelings. We can use jargon or magical thinking to convey our needs, making it hard for other people to perceive what we really need; instead, they hear a criticism or how they may be failing us.

Not taking responsibility: Mary, what you said really bothered me. I need a friend right now, and you're not being one.

Taking responsibility: What I need right now is for someone to hear how I am actually experiencing this change.

Making requests. NVC requests answers to two questions that lead to a joyful and compassionate life: What contributes to our well-being? What would make life more wonderful? Once we can attend to what we observe, feel, and need, we become clear about what specific requests we can make so that others can respond compassionately. That means clearly asking for what we *do want* rather than what we don't want—using positive rather than negative phrasing.

> *An unclear request*: I don't want you to reject me or my son. (This could be stated positively but still be vague: I want you to accept us for who we are.)
>
> *A clear request*: I want you to tell me that you value our friendship even if my family does some things you don't agree with. Maybe you would be willing to hear me present to parents at the next LGBTQ talk?
>
> *Reflecting clarity*: Was that clear? ... Am I making sense to you? ... How did you understand what I said? (A simple reflection question framed positively can help everyone reach understanding.)

Remember, people are doing the best that they can with the skills that they have. Being mindful of how we communicate is an important first step. If you are curious to learn more about compassionate communication, you can explore these skills at the Center for Nonviolent Communication.

When we wake up from automatic habits of thoughts and beliefs that are rooted in judgment, we begin to speak from the heart and we begin to relate with kindfulness. We learn to welcome our own and each other's stammering or silences with more conscious speech. An internal reckoning takes place. We find it takes time. As we bring compassionate communication into our awareness, we wonder:

What am I reacting to?

What am I feeling in this moment?

If I dig deeper, what needs are connected to what I'm feeling?

Am I making a clear request in a positive way?

Asking these questions marks a turning point: the SPEL is broken and we begin to experience the ingredients of PEPPIE. As Elyse points out, "As humans we need to all do a better job at recognizing our judgments and conditioning, take the extra step to realize there are many paths that lead to the same place and that it's not our job to decide someone else's journey." By being present, managing emotions, and keeping perspective, fertile ground is laid for compassionate understanding.

Reflection: *When I am clear on the needs that underlie my feelings, I can more readily show compassion to myself and others, find common ground, and allow people to be as they are.*

CHAPTER 19

Singing for Our Souls

Singing together can be filled with delight. My mother loves to describe how, at three years old, we would constantly croon a sad love song about a dying rose in German: "Heideröslein," or "Little Rose of the Field." I was so happy to sing. As I grew older, self-consciousness took over and I became shy, anxious, afraid to speak up—and terrified to sing in public. When I failed to make the chorus in sixth grade, I stopped singing completely. In high school, while my friends were belting out Journey, Springsteen, and U2 lyrics, I faked it by humming along. Even when my own teenage daughters went retro and belted out "Don't Stop Believin'," I silently cringed as if the universe were making fun of this small-town girl.

Then I attended "shame camp." That's our affectionate label for The Daring Way training program by Brené Brown. To help others build shame resilience and learn how to "show up, be seen, and live braver lives," we future facilitators looked squarely at our own shame for three days straight. The icebreaker was a karaoke-style sing-along in German to "Ode to Joy" from Beethoven's *Symphony No. 9*.

Singing? In German? I was put right in a vulnerability mode and there was nowhere to hide. And yet, in those few minutes of singing, a communal bond among a hundred people formed. I could barely sing through my tears. Something important shifted for me in this room of strangers. No one had been judging me but me.

When I returned home from the workshop, I made a bold move to join a singing circle called Woman Song. In this safe space, women learn music by ear and by repetition—no experience or music reading required. We just show up. I hummed along for weeks and weeks until lyrics began to stick. Never was the power of sound so revelatory than in this company of caring women. We sang for ourselves, for each other, and at assisted living homes as acts of service. For me, joining this group was a small act of bravery. If ever there was a receptive, nonjudgmental audience, it was there. I joined a chorus after all.

Making Music Together

As Emily Dickinson wrote, "Hope is the thing with feathers that perches in the soul and sings the tune without the words and never stops at all." Singing is emotionally transformative and strengthens social bonds. It has historically persisted across generations and cultures, which suggests that it is evolutionarily adaptive. The first sounds we ever hear are lullabies—rhythms of comfort—which feed our basic need for social contact. We learn the alphabet through group singing, and we mark many traditions, rituals, and celebrations with singing.

"Laughter, song, and dance create emotional and spiritual connection; they remind us of the one thing that truly matters when we are searching for comfort, celebration, inspiration, or healing: We are not alone," writes Brown in *The Gifts of Imperfection*. In this way, music can lead to "co-pathy," the social aspect of empathy that arises when people in a group emotionally resonate with one another. This interpersonal resonance can lead to less social conflict, greater group cohesion, generosity, and healing. It's an ensemble of kindfulness. When we sing together, we are better together.

Many Hearts Beat as One

Your experiences of music and, in particular, singing have deep neurological correlates. According to Stephen Porges, the frequency of many of our favorite instrumental melodies duplicates the band of the human voice, called "prosody," and has the same frequency that mothers use when singing lullabies to babies. These melodic vocalizations, and the facial expressions you make when singing, trigger the neural mechanisms of your self-soothing system.

Music is simply good for your health. A meta-analysis of more than four hundred studies on the physical and psychological benefits of music interventions concluded that playing and listening to music enhances the immune system and reduces levels of stress, stimulates anti-inflammatory properties, and is more effective than antianxiety medication in lowering anxiety before surgery. Music therapies are known to help people cope with trauma and loss.

Even sad music that makes you cry has a biochemical reward, resulting in nostalgia, serenity, or relief. No wonder I loved those melancholic German lullabies: sad music has a paradoxical effect on well-being, as feeling moved by music is related to dispositional empathy—your ability to imagine and even experience what another person is going through.

In a music-listening study of sixth-graders, scientists looked at the relationship between empathy and sad music. Children with higher scores on the fantasy dimension of empathy, or the ability to imagine oneself in another's situation, preferred sad music. These children experienced the pleasurable aspect of emotional arousal, the "sweet emotion," because while the music was sad, they experienced positive emotions. The same study showed that children attending weekly one-hour music lessons throughout a school year significantly increased empathy scores compared with children in two control classes.

No matter what kind of music you listen to, play, or sing, it will have social benefits. Your compassion network is activated, enabling you to feel connected with others. Neurochemically, hearing music releases oxytocin and vasopressin, known to regulate social affiliation and bonding. Happy songs, such as "Walking on Sunshine" or "Brown-Eyed Girl," lead people to be more cooperative, regardless of whether the music boosts mood.

Scientists who study the effects of music on the brain and body have helped us understand why singing together can move us profoundly. When singers are in harmony, so are their heart rates, and a physiological synchronization occurs as singers' hearts and lungs oscillate at the same frequency. Synchronized musical activities, such as group singing, chanting, dancing, and drumming, reflect a kind of social coordination and promote feelings of trust and well-being. Simple mantras, sound vibrations, or short phrases of affirmation have been sung in social and religious groups for centuries. Girish, international musician and author of *Music and Mantras*, writes, "Our singing voice is a direct, living connection to the deepest parts of who we are, and finding that voice naturally heals, aligns, and empowers us." In effect, your brain and body rhythms sync up within you and with others—and you get a neurochemical happiness cocktail.

Kindness in Practice: *Strike a Chord*

Zora Neale Hurston wrote, "Love, I find, is like singing. Everybody can do enough to satisfy themselves though it might not impress the neighbors as being very much."

> For thirty minutes a day, intentionally bring music into your life and savor it. Sing in the shower or car. Even better, sing with others. Pick up the instrument that's been collecting dust for years, or try a new one. Find opportunities to sing in religious services, holiday gatherings, or family sing-alongs. You can also create song playlists that inspire or comfort. Take them with you as you walk or run outdoors, drive, cook, or simply sit and listen.

There may be nothing quite like the sound of music to evoke kindfulness and reach deep into our souls. Singing requires deep breathing, which grounds you in the present moment, triggering those core PEPPIE ingredients of *Presence*, *Emotional regulation*, and *Perspective*. In the whirl of an anxious world, you need to reset your overactive mind and anchor your racing heart. Song and dance illuminate our primal connection to one another, just like nature can bring us down to earth and the soft touch of a hand can offer more comfort than words. These are expressions of harmony, hope, and healing—and they can fill your heart with profound kindness, restoring you and others to wholeness. Being kind is life's love song that we sing to one another.

Reflection: *When I sing and dance, my body harmonizes with life.*

CHAPTER 20

Cherishing the Little Things

When it comes to love and romance, we may set such lofty ideals for someone that we miss the small moments that actually bond couples together. "Comparison is the thief of happiness," notes Brené Brown. Lisa was definitely tied to a mental checklist. She was looking for the great-looking guy, the one with the most wealth, the charming dude with an impressive résumé. She wanted a Prince Charming to ride in on a white horse and sweep her away. Instead, she found herself mistreated and humiliated by shallow men.

It's a common story. In her version, Lisa habitually gave away her power over her own life to these men. There was no commitment, no "I love you." Then something that had nothing to do with Lisa's love life—or so she thought—happened. Lisa's father swiftly passed away from cancer, and Lisa became her mom's companion. "We had a standing date. We would do Pilates together every Friday. We would make dinner together. We spent a lot of time in the house together, watching old movies. I learned so much about my parent's marriage and how strong their bond was. I remember Mom telling me that they would hold hands as they fell asleep every night. For two years, I was repeatedly moved by her stories of the little things they did together." Her parents were true friends.

Shockingly, just two years after her father's death, her mom was diagnosed with advanced cancer. It was a one-two punch for Lisa. "There was this moment when Mom was dying and we went to the emergency room together. She was on painkillers, and they said to her, 'What year is it?' She said 1965. That was the year Mom and Dad met. I thought, *Oh, she's going back. She wants to be with him; she's going back there.*"

After her mother died, it took Lisa a long time to recover from her losses. "It was like walking around with an open gunshot wound. I became clearer in what I wanted after that. Everything—all the myths about finding Prince Charming, everything that I thought love was in a relationship—was no longer." Lisa reflected, "Even though I never want to relive it, I will always be grateful that I had that time with Mom. It was a gift: *I learned how to love.*"

Eventually, Lisa did meet someone and marry, and it was a small moment that fused their bond. "One day he told me a story about his mother. She had collapsed at a wedding reception—and died in his arms. It was heartbreaking. I thought, *Oh my God, this is a connection that I didn't know we had.* When he shared his story with me, I then shared mine with him." A friendship was sealed.

Long ago, Lisa wrote a list of qualities she wanted in a partner. The first word was "kind" and the last quality was "will cherish me." To cherish someone means to care for and protect someone lovingly—to hold them dear. Lisa had always known what she needed but didn't know how to find it until she understood more about the day-to-day expressions of love in her parents' partnership.

Microscopic Moments of Love

The author Cheryl Strayed writes: "The healing power of even the most microscopic exchange with someone who knows in a flash precisely what you're talking about because he or she experienced that thing too cannot be overestimated." Strayed's soulful book *Tiny Beautiful Things: Advice on*

Love and Life from Dear Sugar offers a raw and simple remedy for connection: Be loving. Show it. Say it.

To one reader afraid to say "I love you," she wrote, "Do it… Don't be strategic or coy. Strategic and coy are for jackasses. Be brave. Be authentic. Practice saying the word 'love' to people you love so when it matters the most to say it, you will." Three little words can have huge effects, especially when you mean them and repeat them often. Practice is the key word, and the earlier we start in life the better.

Andy Gonzalez, cofounder of the Holistic Life Foundation, agrees. He teaches mindfulness to inner-city youth, young women living in homeless shelters, and large audiences of entrepreneurs. He thinks that saying "I love you" should not be limited to people we feel romantic toward. That's too narrow a definition, insists Gonzalez, who believes the whole of humanity is deserving of hearing those three special words. He leads by example, telling the kids he works with "I love you" so often that eventually the kids start saying it. His approach challenges people's judgments. "You can see the kids thinking, 'I'm supposed to love the people I don't like?' Then we have conversations with them and ask, 'Don't you think those are the people who really need love?' They're the ones who need a lot extra.

"When you're always exuding love from the inside out, it spreads," says Gonzalez. "There's a thing we always say: 'We're like love zombies.' It's like we're infecting people with love. That's the goal." This positive contagion effect is real. Emotion researcher Barbara Fredrickson broadly defines love as the master positive emotion that results in a kind of "positivity resonance" in long-lasting relationships and friendships. It also occurs in micromoments among strangers. Love encompasses any positive emotion in her list of ten—joy, gratitude, serenity, interest, hope, pride, amusement, inspiration, awe, and love—shared between two or more people. Love, she says, is the "all of the above" positive emotion, an energizing blend of positivity, interpersonal synchrony, and mutual care. To me, this affirms that kindness truly is love in action.

Bids for Connection

Grand gestures don't seal our bonds; everyday positive interactions do. This was Lisa's slow awakening about her parents and her own marriage. Psychologists John and Julie Gottman have been studying couples for forty years in the Love Lab at the University of Washington. While couples interact in a mock living set over a weekend, researchers study them by analyzing video recordings, taking physiological measurements, and following up with them years later. They concluded that the secret to stable and lasting relationships is a mathematical equation: a magic 5-to-1 ratio. As long as there are five times as many positive interactions between partners as there are negative ones, a relationship is likely to be happy and enduring. The Gottmans can even predict whether newlyweds will stay married based on the first three minutes of a fifteen-minute interaction. They look at marriage from a unique lens and have ways to identify the "masters" versus "disasters" when it comes to marital bliss. It's not a matter of which couples argue or fight more or less, as one might assume. Rather, master couples are calmer, more flexible, and more forgiving; cultivate trust and intimacy; and acknowledge positive things about their partners. Disaster couples show patterns of mistrust, scan for negative characteristics in their partners, and are prepared to fight or flee. In other words, it is not the words that matter most but subtle and unspoken qualities in the interactions: touching, smiling, laughing, and giving compliments. I think of Lisa's parents holding hands in bed every night.

The simple truth about happy relationships, writes John Gottman in *The Seven Principles of Making Marriage Work,* is deep friendship. "These couples tend to know each other intimately—they are well versed in each other's likes, dislikes, personality quirks, hopes, and dreams. They have an abiding regard for each other and express this fondness not just in the big ways but through small gestures day in and day out." When couples connect, or make "bids" for connection, they *turn toward* one another with admiration, attention, and acknowledgment of each other's needs. They do kind things. Real love, as Sharon Salzberg says, is more than an emotion; it is an ability.

Even when there are disagreements, happy couples are in a state of "positive sentiment override," in which they work out problems in a spirit of mutual caring and positive feelings. Julie Gottman states, "Kindness doesn't mean that we don't express our anger, but kindness informs how we choose to express the anger. You can throw spears at your partner, or you can explain why you are hurt and angry, and that is the kinder path."

In her excellent summary of Gottman's research, and one of the top-ten viewed articles in *The Atlantic*, Emily Esfahani Smith concluded that the secret to love is just kindness. As she explains, "There are two ways to think about kindness. You can think about it as a fixed trait: either you have it or you don't. Or you could think of kindness as a muscle. In some people, that muscle is naturally stronger than in others, but it can grow stronger in everyone with exercise. Masters tend to think about kindness as a muscle. They know that they have to exercise it to keep it in shape. They know, in other words, that a good relationship requires sustained hard work." Now that's joyful *effort* at work.

It makes sense that master couples are healthier and live longer. They show lower physiological arousal when stressed, whereas the disaster couples evince high fight-or-flight arousal during their interactions, even if they aren't arguing. Happy couples say "I love you," they kiss and touch nonerotically every day, and 88 percent of them have a date night every week. It's the little things that really count, the daily acts of cherishing life together. This is good to know for all relationships, not just the intimate ones.

Kindness in Practice: *To Have and to Hold*

It seems so simple to acknowledge what we cherish. Yet we can easily take such things for granted. A vital part of cherishing is scanning for these tiny, beautiful moments and appreciating them. To do so, it's important to be present and make efforts to connect, whether keeping up to date and curious about someone's daily life or supporting future dreams. Here is an adaptation of the "cherishing" skill from The Gottman Institute, which is dedicated to research and training in fostering healthy relationships. Use it as an ingredient of joyful effort in enhancing any relationship you care about.

> In your journal, write qualities that you cherish in your partner, friend, family member, or child. You might jot down characteristics such as curious, determined, funny, loyal, witty—the possibilities are numerous.
>
> Next, write this person a note of adoration or appreciation, using the words you chose. Express your cherishing by describing why, and remember to include small, quirky reasons. End the letter with statements of love.

It's funny that the science of love proves the obvious. Because we're wired for love and kindness, any avoidance, criticism, or contempt corrodes a bond, wears down the body's defenses, and even shortens lives. We can always love each other more, be kinder, and speak our hearts. We can even love someone so much it that cracks us wide open.

Up until now, you have read about experiences that can get in the way of your instinctual capacity for love and kindness, unwittingly catching you in a SPEL. Breaking the trance means exposing yourself to all of life: good, bad, ugly, and beautiful. It means trusting your body to be a messenger and that your inner gifts can respond with love and kindness. Kindfulness is a daily practice. When you open up to the wholeness of life, you can appreciate the struggles and triumphs. In doing so, you pave the way for one of the most gracious ways of kindfulness: gratitude.

Reflection: *I commit to cherishing small moments of love and kindness in my relationships, knowing there is no better time than the present.*

CHAPTER 21

An Attitude of Gratitude

Kindness and gratitude go hand in hand, as we both receive and express sentiments of appreciation. *Kind World* radio series producer Erika Lantz has noticed this as well: "A big motivator for people telling their stories is that they wish they could adequately thank the person who has had an impact on them. This is one way for them to do that. They want to let the world know, 'Someone did this for me.' It's a way of recognizing those who helped."

Sometimes we simply aren't sure how to express appreciation. Over the years, many people have shared with me their stories about acts of gratitude. Here are a few true tales, in the words of those individuals, from my collection of "kindness dispatches" that convey the power of gratitude:

- Once I found an anonymous note attached to my car. It read, "Thank you for your kindness and your smile. You don't know how much it means to me." I never knew who gave me that note, but it had a profound effect on me.
- I grew up in the projects without a lot of money. My first writing tutor bought me a word processor, back in 1995. Because of that

kindness, I am a published award-winning playwright now. I'll never forget what he did to get me going.

- I was in a cab heading to the airport to catch a flight to New York, when I realized I had forgotten my wallet. This was in the days when you did not need an ID to fly, but I did need cash for cabs on both ends. I was going to a job interview and could not cancel! When I explained this to the cabbie, he said, "I know what it's like to be out of work, Miss." He handed me $200 to cover cabs in NYC. I was so thankful. The next day I gave him back $300. He had no idea that would happen. He was just a lovely and generous man.
- I love sending cards to people, sometimes for no reason. I just let them know they are on my mind and that I miss them. It gives me so much joy. After the school tragedy in my town, Sandy Hook, Connecticut, we all started leaving notes for each other and the people who helped us.
- During my cancer treatment, a coworker donated her vacation hours so I could go through surgery without worrying about missed pay. I hope she knows how grateful and humbled I am by that thoughtful and selfless act.
- When my brother died forty years ago, I was given a book of comforting quotes by a friend. I still read it. This friend was not close, and we lost touch, but I'll be forever grateful. Small kindnesses like this keep me going far more than grand gestures.
- My daughter was given a scholarship for college. She deserved it for her grades, effort, and character. The special family foundation that chose her knew we needed every penny. They could have selected another child, and they could have kept the money. Instead, they were kind. They helped us. I still tear up thinking about it. It changed our lives.

Ordinary Blessings

"If the only prayer you ever say in your entire life is 'thank you,' it will be enough," said the theologian and mystic Meister Eckhart. Countless sages and spiritual masters, contemplative leaders and writers, all emphasize how essential gratitude is in a meaningful life. Going far beyond good manners, true gratitude comes from the openness of deeply understanding the fragility and beauty in life. This leads to acknowledging blessings in ordinary experiences. It is a practice of appreciating.

David Steindl-Rast, a Benedictine monk, writes, "Gratitude springs from an insight, a recognition, that something good has come to me from another person, that it is freely given to me, and meant as a favor. And the moment this recognition dawns on me, gratitude too spontaneously dawns in my heart: *Je suis reconnaissant*—I recognize, I acknowledge, I am grateful; in French these three concepts are expressed by one term." In this translation, Brother Steindl-Rast offers the secret formula for experiencing gratitude, and it all begins with recognizing the good.

Effects of Being Thankful

Gratitude is an inclination toward noticing and appreciating the positives in life. In this way, gratitude is both a feeling and a life orientation. It encompasses good deeds and also tangible and intangible things, such as a rosy sunset, medicine that eases an ailment, or the freedom to vote.

Gratitude has evolutionary underpinnings and is part of our caring blueprint. Scientists believe that expressions of gratitude are fundamental for social relationships in ways that nudge humanity from one generation to the next. It's an "I'll scratch your back if you scratch mine" sentiment seen in other primates. This social exchange is known as *reciprocal altruism*, which we experience when we offer a helping hand to someone and he or she does something kind in turn. Both people are left with positive feelings, making it more likely that the exchange will continue. Sociologist Georg Simmel called gratitude "the moral memory of mankind."

This exchange creates an upward spiral of well-being: studies show that in addition to being closely associated with positive feelings such as happiness, gratitude plays a role in personal well-being, self-acceptance, having purpose in life, post-traumatic growth, and physical health. Typically, gratitude studies ask people to count their blessings, write in gratitude journals, or hand-deliver a heartfelt thank-you letter. They are then compared with people who did something else that required a similar level of effort. It turns out that our whole being is affected by gratitude—psychological, physical, and social—and these positive effects last for a while. Gratitude boosts happiness and life satisfaction, improves sleep, strengthens relationships, inspires forgiveness, promotes altruism and helping behaviors, helps people face adversity including death, and may even directly impact heart health in heart-failure patients.

Leading scientific expert on gratitude Robert Emmons suggests that there are four main ways gratitude contributes to our lives. Being grateful:

- Magnifies the goodness in life through celebrating or appreciating the positive in the present moment;
- Blocks out negative emotions, such as envy, regret, and resentment, because you can't be grateful and resentful at the same time;
- Buffers stresses and speeds up recovery from physical symptoms and trauma, in part by reinterpreting life events from a new angle—like seeing a silver lining; and
- Enhances a sense of self-worth by feeling cared for or appreciated by someone else.

It's true that sometimes instead of gratitude we feel guilt and indebtedness: others can give so much to us that we worry we'll never be able to repay them; or deep inside, we feel unworthy of assistance; or we feel bound by obligation. One study found that what makes the difference between indebtedness and gratitude is the extent to which we focus on the benefits to ourselves.

Basically, if you value being an independent and self-made person, it will be harder to experience true gratitude because receiving help goes against your grain. The same effect occurs when your primary reason for being in a relationship is to benefit yourself, to gain safety and security for instance, rather than to nurture another. When you see benefits in mutual nurturing, base connections on that, and stay as attuned to other people's happiness as to your own, you will swing toward gratitude.

Kindness in Practice: *Recognize, Acknowledge, and Be Grateful*

Because positive emotions such as gratitude are fleeting (as are all emotions, for that matter), cultivating an attitude of gratitude goes far. Following Rick Hanson's notion of "taking in the good" is a great way to practice gratitude so that the good things stick with you. In chapter 14, "Taking in Kindness," I described Hanson's three-step process that trains the neuroplasticity in your brain to overcome the negativity bias and become more positive. To cultivate gratitude, simply add a fourth step:

1. Notice or create a beneficial experience.
2. Be present.
3. Let the experience stick in your mind.
4. As you appreciate the beneficial experience, say "Thank you."

This can become a simple, quiet, inner habit as you go through your days. You could be thanking anyone or anything: God, the universe, Lady Luck, or nature. If someone did something kind for you, start with feeling gratitude. That way, you will cultivate the true benefits of gratitude, and any gesture you make to express it will come from the heart.

"Who can distinguish giver and receiver in the final kiss of gratitude?" asks Brother Steindl-Rast. Gratitude gives meaning to daily life, allowing you to feel that your participation in the world matters and that we belong to each other. When you pause long enough to behold the wonders before you, appreciate kind gestures that delight or surprise you, or acknowledge the smallest comforts and joys, you change that much more for good. Integrating kindness and compassion through gratitude bolsters your caring circuitry, ignites energy, and builds resilience. Your lens widens on the inevitable joys and sufferings, with a view of love. That's what I call *kindsight.*

Reflection: *As I make efforts to feel grateful, I become more aware of the delightful and surprising moments in my life.*

PART 4

Kindsight

noun kind · sight\ \ ‘kīn(d)-sīt \

: viewing life experiences with tenderness and understanding

CHAPTER 22

Grace and Grit

I first met Kriss in a woman's trauma group that I co-led many years ago. She was the youngest of the group, yet her emotional intelligence was strong, her manner quiet and gentle, and she had a keen empathy with everyone. This was remarkable given that she experienced a childhood of abuse and neglect with many adverse childhood experiences (ACEs), as described in chapter 7.

Kriss had undergone a number of psychiatric treatments, some that did more harm than healing. When her therapist moved away, she asked to work with me privately. Over time I became witness to her inner warriors and protectors, the meek and vulnerable parts, and the destructive forces too. Many of these personalities were captured in journals and voicemail messages as she used the power of her empathic imagination to protect and heal. But her struggles ran deep, and she battled afflictions from toxic stress, including anxiety, chronic fatigue, and chemical sensitivities. It was a wonder Kriss could get up every day. And sometimes she couldn't.

At times, I was caught under my own SPEL. I felt overwhelmed with supporting Kriss's many needs; I had my own moments of empathy fatigue, lack of confidence, and the helplessness that can arise during trauma work. In the middle of one night, I received a page and saw her number appear. I called her back, anticipating she was having a panic attack or nightmare.

"No, I didn't page you," Kriss said. "Are *you* okay?"

I doubled-checked my pager and there was no phone number. I had been in my own bad dream. When I next saw Kriss, I apologized for disturbing her sleep, riddled with my own embarrassment. Quick as a whip she countered, "That's okay. Now I know you're human."

We worked determinedly toward her goal of attending college. She slowly put one foot in front of the other as she sought help from "worthy and safe guides along the way." One day, Kriss came to her session clutching her college application packet. It was a momentous occasion. We silently walked to the blue mailbox on the street corner. Together, with both our hands on the package, we let it drop in. We heard the thud. Then we turned to each other with a smile and a shrug to the universal powers that be—it was in their hands now.

Years later, Kriss recalled that moment as a shift from being in survival mode to seeing a glimmer of a good future. "It was like I could see the shore and believe it was possible to get there."

Kindsight

When you experience moments that change how you look at the world, how you see the past, and how you imagine the future, kindsight arises. It is born of an awakening that occurs in the present moment, similar to a vast landscape coming into view. *Kindsight is your ability to view experiences from a place of tenderness and understanding.* Sometimes kindsight reveals a spontaneous lesson that helps you flourish. At other times, it unfolds slowly over many years as you commit to learning from hardship or choose to take beneficial risks. And it can also be experienced as a humble faith that somehow, someway, your life has purpose. Kriss came to her own recognition: "You will fall on your knees a thousand times. Yet, every moment is an opportunity to listen to your fear. And it's also an opportunity to latch on to hope."

The stories that follow in part 4 reveal experiences of *kindsight*—reckonings born from the darker experience of shame, the fragility of being aware of flaws, feelings of being wronged by others, and the myriad challenges of being human. Kindsight fosters the empathy we need to recover from our failures, the forgiveness to repair our relationships, and the compassion to connect us with a more nuanced understanding of the human experience.

Krista Tippett, host of the radio show *On Being*, links the compassion we can have toward ourselves with its ultimate power: "What a liberating thing to realize that our problems, in fact, are probably our richest sources for rising to this ultimate virtue of compassion, towards bringing compassion towards the suffering and joys of others." Kindsight marks such a deepening and expanding. It helps us integrate the whole of our lives with kindness.

Creating Upward Spirals

Like Kriss, you can heal, transform fear into faith, find hope, and imagine new possibilities. As part 3 showed, you can purposely create positive experiences full of kindness. You can ignite an upward spiral for positive change. This momentum gives you a new view—a kindsight—for reenvisioning previously held, narrower views about your life so that you no longer feel like a victim of circumstances or like a passive bystander as the world turns before you. You can learn to love the harmful, ashamed, or destructive parts as inner protectors and guardians worthy of respect and care. These aspects can be relieved of their duties, allowing you room for new possibilities, passions, and goals—which in turn inspire more and more positivity.

When you experience just a moment of positive emotion, a window in your mind cracks open to new possibilities. It is a breath of fresh air that, quite simply, opens you up. It's true that positive feelings such as amusement, calm, courage, joy, and serenity can feel fleeting. Even so, they invite in new ideas, inspire creativity, allow you to reinterpret situations, evoke reflective insights, and drive you toward reconciliation.

Psychologist Barbara Fredrickson has a theory that illuminates a path toward healing and joy. As mentioned in chapter 1, with positive emotions you gain the ability to broaden and build your inner resources. Some of this happens because positive emotions are adaptive—they help you become more able to meet the conditions of your environment in healthy ways. So they are good for survival because they build resilience. This happens one joyful, if ephemeral, emotion at a time, and as more time passes you can see a panorama of potential in your life. As you find yourself standing in this new landscape of awareness, you become an agent of change, a protagonist in a new and original story. These experiences generate even more positivity. She calls this the "undo effect" enabled by an upward spiral of momentary positive emotions.

Fredrickson and her colleagues designed studies in her Positive Emotions and Psychophysiology (PEP) Laboratory to test her theory. In one workplace study with employees at a technology firm, the PEP team tested the potential of inducing a steady "diet" of positive emotions though meditation, specifically a loving-kindness meditation (LKM) like the one described in chapter 5. Half of the employees in the study received six one-hour weekly meditation instructions and were given recordings to listen to at home, while the other half were on a waitlist. The results showed incremental yet increasing shifts in positive emotions, greater mindfulness, and decreases in reported physical symptoms such as headaches.

Moreover, those people who showed significant increases in positive emotions from the LKM practice during the study were still at it more than a year later. They adopted a new and pleasant habit of mind. Later PEP studies looked at individual differences after LKM training based on more objective data, such as physiological arousal measured by cardiac vagal tone, or automatic facial response such as eye gaze and a genuine smile, with similar beneficial results. Even less-intensive practices can boost positive feelings, such as reflecting for a few minutes on how close or in tune you feel about your enduring relationships, or noticing what you appreciate in your day.

The upshot is that as positive emotions accrue over time, your well-being improves and durability sets in. So purposely prescribing positivity in your life will have physical benefits as well as psychological ones, such as expanded awareness and curiosity, a greater sense of social connectedness, and, importantly, an ability to "savor future events." Even with the most tender of steps forward, my patient Kriss cultivated a sense of hope and optimism about her future.

A large body of work now shows that people who are able to "bounce back" from stressful or negative experiences are able to face hardships in effective and adaptive ways; and those who stick with positivity practices are happier and healthier. They also have tenacity, or "grit." According to psychologist Angela Duckworth, *grit* is the result of perseverance and passion for a particular long-term goal, just like Kriss's goal to get into college in spite of her setbacks. Having an ability to view stressful situations as challenges rather than threats increases the likelihood you will stick to achieving a goal, even if it may seem far away. If you want the resilience to follow your dreams, it helps to see the positive meaning in negative experiences and failures along the way. To say it another way: make lemonade out of lemons. This kindsight gives you a long view: if the going gets rough, you'll survive disappointment and get back on your feet.

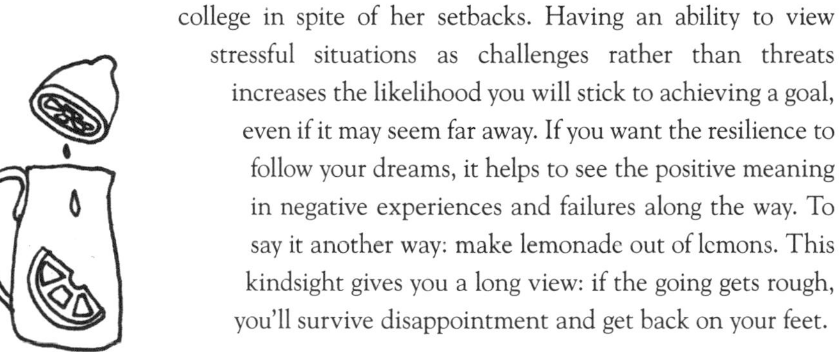

Kindness in Practice: *Enhancing Positive Moments*

The good news is that you can cultivate grace and grit by creating micro-moments of positive emotions. They will open your heart to allow kindness and compassion to spiral from the inside out. The practice of both *noticing* the experience of positive emotions and *cultivating* them is as essential as taking a daily vitamin, exercising, eating well, and having quality connections with others. And your positivity begets positivity.

The ten positive emotions Fredrickson studies are joy, gratitude, serenity, interest, hope, pride, amusement, inspiration, awe, and love. You can notice these and others. Go to the list of emotions in chapter 3 and experiment with expanding your awareness of the positive feelings in your life. Expand your view of the past with them in mind, and envision a future filled with them. See your mistakes as failing *forward* with new insights along the way. Ask yourself how you might build up incremental moments of positivity as you cultivate a daily practice of kindsight.

A simple way is to check in with yourself every day for a month. Reflect on a moment, no matter how brief, and describe the feelings that passed through you. Be all in.

> Today I felt [positive emotion] while [describe interaction/situation]. As I reflect on this moment of positivity, I notice that [describe an insight or new perspective or intention].

> Today I felt [amusement] while [in line at the coffee shop. The lady in front of me forgot her wallet and was so upset. So I told her it was on me. She started to cry with relief, and that made me laugh because it was no big deal. Then she hugged me]. As I reflect on this moment of positivity, I notice that [this little encounter changed my mood for the rest of the day. Work didn't seem as stressful somehow. I don't have to take it so seriously].

> As you engage day to day, notice what happens as you begin to shift your attention toward these little bits of goodness. What might you discover or uncover about yourself and others?

With her creative wisdom and faith in her own healing, Kriss wrote a new story for herself. "Even during moments of great despair, I would ask myself, *If I could go back in time and take a different path and maybe avoid this pain, what would I do differently?* My answer was always the same: I would never choose a different life. There has been a wisdom carved out of the heart of my pain."

Cultivating kindsight pushes your old narratives out to the margins on the page—and with messy scribbles, notes, and strike-throughs, you can develop a new story. Doing this allows the painful parts to heal and new parts to thrive. It also allows you to rewrite the versions of your life told to you by others, a process that sparks personal agency and illuminates core values. Kindsight ushers in a broader view of life; it makes room for inner strengths and invites compassion for yourself and others to emerge. Ultimately, kindsight connects you to love.

Reflection: *Through kindsight I can reflect on my experiences from a place of tenderness and understanding.*

CHAPTER 23

Forgiveness Is a Gift to Yourself

If you've been badly wronged, whether by parent, partner, or friend, your first reaction was probably instinctual. You may have wanted that person to pay, and you might have sought revenge or avoided the offender altogether. And the chances are good that you have carried that hurt or spirit of vengeance ever since. That's what I chose to do with my father for much of my adulthood.

When I was ten years old and my sister was nine, my dad picked us up on a brisk, bright Christmas morning, just as he had done ever since my parents' divorce. The plan was to spend the day opening up presents from Santa and eating good food. But when we hopped into the car, he told us he had a special surprise. We were going to take a trip, a *real* trip.

During the hour-long drive to the airport, my thoughts raced. I knew Mom would never let us go on an overnight trip anywhere without a well-packed rucksack of snacks, toothbrushes, our favorite stuffed animals, and the most crucial item: my sister's security blanket. I realized Dad was once again scheming to get back at my mother for having the guts to leave him. I wondered if I could stop him with my tantrum superpower, but we were racing along the highway and there were no stoplights where I could jump out of the car and force the issue. I went into a dissociated panic, frozen; I was onto my dad and he knew it. We locked eyes in the rearview mirror for a long moment, and I felt helpless to do anything to stop his cruelty toward my mom.

For weeks, we visited distant relatives, trapped in my father's vindictive dog-and-pony show of fatherhood. The "Christmas kidnapping," as I came to call it, was a turning point for me, my sister, and my mom. Each of us snapped in some way. Upon our return, my mother melted into our bodies at the first embrace, her relief so palpable it took my breath away. She was never quite the same, and her overprotection suffocated us to the point that, as teenagers, we rebelled. I couldn't ever blame my mother. But it took me a really long time to forgive my father.

Freedom in Forgiveness

With past pains, a question inevitably comes up: How do you forgive someone who has really hurt or betrayed you? Sometimes it feels impossible. But the alternatives are grim. Clinging to resentment, anger, or vengeful fantasies leaves you with an uncomfortable paradox: keeping a painful story alive means you continue to give your power away to the people who hurt you. Yet deciding to forgive and to move on can feel like the destruction of your identity as you have known it—a mini death, scary and raw. And that's what is at the heart of our resistance to forgiving: a fear of being broken into a million pieces.

But here's the thing. That resistance is a ping on your incredibly intelligent emotional radar system. It's telling you that something must change. It's pointing out that new awareness is possible, but you need to *go toward it*—or, as Byron Katie says, "If you want to find a way out, go in."

Forgiveness, like compassion itself, is a practice of turning toward the unwanted stuff so that you can transform it into something beautiful. This compassionate engagement with past hurts is a powerful use of kindsight. By courageously reimagining who you are and who you want to be *right now*—informed by your past but not imprisoned by it—you can find freedom. Because the truth is, deep under the protective identity you have been afraid to release is your unbreakable and beautiful soul. Your inner light is trying to shine through. It needs your love and kindness. It needs to forgive.

Heal Yourself with Forgiveness

Forgiveness is not something you do because others say you ought to. Messages of "just get over it" or "move on" can make us feel invalidated, infuriated, and ashamed. In fact, there are a lot of unhelpful misconceptions about forgiveness: that it requires letting hurtful people off the hook, keeping them in your life, or letting them back in. That it means forgetting what happened or stopping how you feel. None of these things are essential to forgiveness, though some of them might result from it. In fact, forgiveness doesn't even mean you need to tell someone you forgive him or her. Because forgiveness is not something you do for another person. Rather, it is the kindest possible thing you can do for *yourself*.

When we're stuck in anger and resentment, ruminating over a past wrong, negative moods build up that lead to chronic stress and misery. We can become so identified with being or having been a victim that our chances for healing, growing, and building healthy relationships get short-circuited. In fact, the difficulties of the past become fused with our present experience as we see our current situation through the lens of the past. Research reveals that people who carry grudges or resentments or who remain unforgiving suffer. Their relationships are harmed, their negative emotions persist, and their health can deteriorate. Everything from life decisions to daily interactions can be affected, contributing to a state of chronic distress. People who have been able to forgive, allowing past hurts to be in the past, have healthier relationships, are happier, and enjoy better health. Forgiveness becomes possible when you see with vivid kindsight that holding on to anger does not serve you.

It also helps to realize that you are not alone. Your situation may be unique, but the experiences of pain, betrayal, rejection, and loneliness are felt by all. People have shown that forgiving is possible no matter the insult: the loss of a child or beloved, abandonment or rejection, infidelity, assault or rape, theft or swindle, a terminal diagnosis or a missed diagnosis. Name the offense and people have experienced it. And so it is only when you can begin to feel safe and held as part of a common humanity that a new story may emerge—one strengthened with kindsight and infused with fresh starts. The reward of forgiveness is our own freedom, integrity, self-compassion, creativity, and joy.

Kindness in Practice: *"You Are Free"*

Fred Luskin, cofounder of the Stanford Forgiveness Project, defines forgiveness in a way that leaves the popular misconceptions out of the picture. It is "the experience of peace or understanding that can be felt in the present moment." Forgiveness is an internal experience and a choice. It takes time and effort. In his work, Luskin has shown that it is a trainable skill.

A favorite teacher of mine and author of *The Lotus and the Lily*, Janet Conner, offers powerful practices that allow us to reframe the whole notion of forgiveness and to build our capacity for it. I share two of hers here:

Untie the knot. This simple exercise calls on the Greek definition of *forgive*, which means "untie the knot." Conner suggests substituting "I forgive" with "I untie the knot." Try it on one of your hurts. See what happens with a simple shift in phrasing.

> "I untie the knot with [name of person]."
>
> "I untie the knot regarding my [hurt emotion such as ***anger/ disappointment***] with [name of person] and choose instead to [helpful action such as ***free/love/appreciate/respect***] myself."

Release your prisoners. This exercise taps your imagination to untie your knots. You can use this visualization as a meditation or as a journaling exercise. Connor's full visualization exercise may be found on her website. Here's my adaptation of it:

> Imagine you are descending into a dungeon. Think of all the gritty details you can conjure up: a damp and dark stairwell, a sweltering compound with little air, and so on.
>
> Face the heavy door that leads into the dungeon. Heave open the door and step inside. Take in the entire scene. Notice the row of cells holding prisoners. How many cells are there?
>
> Go to the first cell and look in. Who are you keeping prisoner? Is it someone who has hurt you or someone you have hurt? You know who it is. Look your prisoner in the eye, unlock the cell door, and motion for your prisoner to leave. There is no discussion, no rehashing of history, no blame, no expectation of an apology or giving one. Simply guide your prisoner out and say, "You are free. You can go now."
>
> Walk to the next cell. Look at the prisoner and then usher this person out. Repeat, "You are free. You can go now." Do this with each prisoner who has been locked in your mind's dungeon.
>
> When all the prisoners have gone, look around. There is one person left. It is you. You have kept yourself a prisoner too. Perhaps for many years and perhaps for the longest of all. Open up this cell and say to your prisoner-self, "You are free. You may go."
>
> When your prison is empty of inhabitants, look around. See the dungeon cells begin to dissolve and the walls crumble. The space fills with warm, loving, healing light. It is transformed from a dungeon of misery to a sanctuary of love. Take in the vast expanse and fresh air. Breathe in deeply and with tenderness say, "Thank you."

This beautiful exercise is something that can be done any time you find yourself captured inside painful stories. You may want to try this with the support of a trusting friend or counselor who can hold a safe space with you.

Anne Lamott has a lovely notion she calls "forgivishness." She writes, "In a fairy tale you often have to leave the place where you have grown comfortable and travel to a fearful place full of pain, and search for what was stolen or confront occupying villains. It takes time for the resulting changes to integrate themselves into the small, funky moments that make up our lives." *Forgivishness* conveys our fits and starts when we confront sources of our pain. When you embrace it, you can bring kindfulness to sensations, emotions, and thoughts about a wound without getting overwhelmed. It allows you to give gentle nudges of compassion to the inner part of you that feels hurt and damaged. This effort toward kindsight makes room for healing moments to happen, such as the following one with my father.

Decades after the Christmas kidnapping, I sat with my father for the last time. In my heart I had made my peace. We crooned along with some old love songs, Patsy Cline and Johnny Cash. After a long silence he said softly, "You girls are the best thing I ever did." It was a small, funky moment. Pure and full of love. And because I had done my inner work to forgive him, I could truly receive this parting gift.

Reflection: *Forgiveness is freedom. As I untie my emotional knots, I open to love and joy.*

CHAPTER 24

Apologies Make for a Friendly World

It's fascinating to ponder the experiences people have growing up and marvel at why some people turn into curmudgeons, others into martyrs, and yet others emerge as diamonds in the rough that radiate raw beauty, light, and love no matter what. When Tim was six years old, he experienced a defining event that helped him become the kind, humble, and humorous person he is.

"I wanted more than anything to be part of a group of neighborhood boys," explains Tim. "Jimmy, Barry, and Gary were the coolest kids around, and I hoped to be a part of their club. I really looked up to them. One day they told me to meet them at their clubhouse because they were going to initiate me into their club. I was so excited! I jumped on my bike the minute school got out. I got there and they immediately tackled me and held me down. One by one, each boy took a turn beating and kicking me. Then they left me there bruised, in both pain and shame."

What went through his mind after the kids took off? In that turning-point moment, Tim decided: "I never, ever want to be like them."

Twelve years later, Tim was crossing a street in his old neighborhood and saw Jimmy on his bike. "He saw me too, hesitated, and then hightailed it out of there. He couldn't even face me. I realized he was still holding on to what he and those kids did to me. But I was free and had moved on." Tim didn't need an apology in order to forgive the boys and move on, taking his life in a positive direction. But it would have been the kind thing for them to do. Of course, had Tim confided in parents, the grown-ups may have forced an apology—only to make matters worse. "What was more important than receiving an apology," says Tim, "was that I've lived my life never being intentionally cruel. It is the choice I made so many years ago."

The Power of Making Amends

"Can true humility and compassion exist in our words and eyes unless we know we too are capable of any act?" asked Saint Francis of Assisi. Compassion widens the lens on humanity: each one of us can be cruel and kind, naughty and nice, hateful and loving, and everything in between. The contrast between suffering and joy wakes us up to life and shows us how to live in the world because, quite simply, being kind makes us feel good. Being mean or causing harm may lead to regret and self-condemnation for some, leaving open the possibility for making amends. For others, unable or unwilling to change, they become prisoners of their own self-hatred.

This is why an apology, when it is authentic and true, taps the wellspring of a compassionate connection between two people. A heartfelt apology helps people reconnect, heals, and inspires forgiveness. It's a difficult thing to say "I'm sorry" and mean it—it requires being vulnerable and brave at the same time. And it can be difficult to accept an apology and find positive ways of moving on.

Being able to apologize, accept an apology, and forgive are powerful. Coming to terms with wrongdoing lets us decide for ourselves what lens we want to look through. Is the world a friendly place or a hostile place? Does the world need our love or our hate? Do we hold on to hurt or can we untie the knot? Albert Einstein said, "The most important decision we make is whether we believe we live in a friendly or hostile universe." And making amends is a core element of a friendly universe.

Offering and Receiving Apologies

Some things are easier to forgive than others. And not all apologies lead to resolution. According to psychologists Ryan Fehr and Michele Gelfand, an effective apology is a nuanced skill that's rarely taught beyond the simple order we hear growing up: "Say you're sorry." Fehr and Gelfand have teased out three important components to apologies:

- **Offers of compensation** are apologies that focus on restoration of equity through an exchange. In Tim's story, if the boys had broken Tim's bike or ripped his sweatshirt, they might have offered to replace the items.
- **Expressions of empathy** are apologies that focus on the relationship and are expressions of warmth, remorse, and compassion. They convey an understanding of the hurt person's point of view and how the offense affected them. For instance, one of the boys might have shared a story about being beaten up by an older brother, apologized because he knew how it felt to be betrayed by someone he looks up to, and admitted it was a mean thing to take it out on Tim.
- **Acknowledgments of violating rules or social norms** admit that a greater expectation for behavior was not met. Most cultures have some version of the "Golden Rule" in which people are expected to treat others in mutually respectful, fair, and kind ways. The boys, who were older, might have said to Tim, "You're littler than us and it was wrong to beat you up just because we are

stronger. We should be protecting kindergarteners, not hurting them. We let everyone down—you, our moms and dads, and the school. We're really sorry."

But here's where the rubber hits the road. People are complicated. We all have different self-concepts and hold different values, needs, and expectations about the world. So even the most eloquent apology can fail when it doesn't match up with the victim's attitudes and beliefs. As you read through this list of descriptions, also consider what conditions you need to be present if you were to accept an apology. Fehr and Gelfand identify three perspectives to consider:

- **People with an independent self-view** tend to be concerned about their personal rights and autonomy. They value competition over cooperation and tend to focus on give-and-take in relationships. Apologies that are offers of compensation or that restore fairness, as in "an eye for an eye," go far with them.
- **People with a relational self-view** value the quality of friendships and partnerships. They care deeply about their connections with others and tend to relationships. Offers of empathy and bids to reestablish connection are the type of apologies that align with them.
- **People who have a greater collective self-view** and group identity value honor and duty. They care deeply about the norms and moral rules of their group. For them, apologies that acknowledge the violation of rules and social norms are the most salient.

Of course, no one falls neatly into a box, but social science affirms the need for a deep listening and understanding of a victim's values and needs when offering an apology. The psychiatrist Aaron Lazare said, "When the apology meets an offended person's needs, he does not have to work at forgiving. Forgiveness comes spontaneously." A victim's needs, Lazare suggests, may include any of the following:

- Reclaiming a sense of dignity
- An affirmation of the shared values which the offense violated
- A clear understanding that the victim was not responsible and is now safe
- Reparative justice through punishment or compensation
- A conversation or opportunity to express grief over losses

Keeping these in mind when both apologizing and considering accepting an apology can help both people to arrive at a resolution. If you have been wronged, ask yourself what you would need for someone to make amends.

Sometimes an apology can't be offered, however. Perhaps the offender has passed away. Or if the offender remains a threat to you, his or her apology may make things worse—and it would be better to keep a distance. And sometimes an offender may show no remorse. This is when the strength in forgiveness is most revealed, because, as the previous chapter showed, forgiveness is ultimately for your own emotional freedom.

Kindness in Practice: *Writing to Make Amends*

Yes, we've all been both the one who needs to hear "I'm sorry" and the one who needs to say it. Acknowledging wrongdoing is an act of kindsight, and so is making amends.

One way to make amends is to first put pen to paper. It may be cliché to say that letter writing is a lost art, but it remains a profound way to think through your deeper intentions and to express yourself fully. Refrain from making amends via a text message or emojis, as the abbreviated form minimizes the significance of what you are saying. A letter's formal structure is respectful. As you write it out, reflect on the components of an apology and what may be needed for acceptance, honoring the other person's needs. Later, after you have gained clarity through writing, you can decide if you want to have an in-person conversation or to send a letter by mail.

Here is an example to get you started. Use your own words, and practice reading the letter aloud when you are finished.

Dear [friend],

This is so hard for me to write. I am really sorry for my [words/deeds/disregard/insensitivity].

I screwed up big time. I am [embarrassed/ashamed/upset] ***about my actions.***

I realize that I [was wrong/blamed you/lost it/didn't respect your position], ***and that is totally unacceptable.***

I can only imagine [the pain/humiliation/rejection/stress] ***you felt.***

I value our [friendship/relationship/collaboration] ***and can't blame you for being*** [upset/angry/disgusted/disappointed] ***with me.***

I hope you will let me [make it up to you/replace/restore/apologize in person].

I realize this may not be possible and that I may already have caused you too much [pain/damage/suffering/embarrassment].

Please forgive me if you can. I will work hard to regain your trust if you let me.

[Sincerely, Warmly, Love,]

[your name]

It is human to make mistakes, and it is human to make amends. Repairing the misdeeds and failures of empathy are some of life's greatest lessons. One of the kindest things we can ever do—for another and for oneself—is to say sorry and really mean it. This is hard. Our confessional culture gives mixed messages about taking responsibility, often glorifying humiliation and shame on reality TV, talk shows, and bully pulpits. An apology, at its best, is an intimate encounter between two people. Rarely is it a quick fix or a tidy conclusion. Apologies take *Effort*.

In this way, kindsight gives a long view to the messy nature of human relationships, giving space to the fits and starts in learning how to trust and belong to one another. Saying sorry may seem simple on the surface, but it is an act of a kind and brave heart.

Refection: *Saying sorry for my mistakes means I honor the needs of other people and am accountable for my actions.*

CHAPTER 25

Writing New Beginnings

Like heartfelt letters of thanks or notes of apologies to others, writing about our own experiences can change us in profound ways. This can be especially so when it comes to our own uncertainty about kindness. When Michael wrote his story of having a sister with the chromosomal disorder Turner's syndrome, he had an opportunity to work it through and realize what he had learned. At age twenty, he had the courage to face his own failings and to awaken to compassion. Here is what he wrote:

> ***My sister and I are best friends now, but it wasn't always that way. As a child, I was teased for multiple reasons. It pains me to say this, but I tended to blame her for the ridicule I received. For quite some time, I chose to not see her for who she was.***
>
> ***When Elisabeth was hospitalized for depression, it was an all-time low for the both of us. I started to realize that I didn't need the approval of others to be happy, and my insecurities shortly dissipated. In the meantime, Elisabeth held not one ounce of resentment toward me. She has a heart of gold. She forgave me time and time again. As the months went by, Elisabeth returned home, and we slowly but surely mended a broken relationship.***

Despite all the daily challenges she has to face, the frequent illnesses, the bottles of medication, the stares from strangers, she maintains an unyielding optimism. For me, she became a representation of the person I wanted to be, a gold standard for a decent human being. I could never be as unconditionally positive as her, but it is something that I continuously strive for. Kindness is not a choice for her; it is a way of life. She has influenced me more than she will ever realize, and for this I will always be grateful to have her as my big sis.

It's wonderful to reflect on kindness, because I have never deeply considered it before. I'd like to think that kindness has become a natural practice. However, sometimes I say or do things out of impulse that I wish I hadn't. If I catch myself, I do my best to learn from my mistake so that it does not happen again. I would also say that kindness weighs into my decision making, and I do try to think *kindly and positively. The hardest part is that it takes a conscious effort to block out irrational and negative thoughts. When I* do *manage to do this, I find that my mood, energy, and overall mindset are greatly improved.*

Michael's story revealed to himself how suffering can be transformed by kindness and compassion. He described in almost textbook fashion the power of PEPPIE in emotional regulation, empathy, self-awareness, and perceptive taking—but all of it arose from his own experience. We generally do not give kindness the attention it deserves, and maybe the reason why so many people are stumped when asked for a story about kindness is because it just isn't on their radar. But as parts 1 and 2 in this book show, it is completely natural to us. And so is telling our stories.

The Power of Story

"We tell ourselves stories in order to live," began the novelist Joan Didion's essay *The White Album*. Her words remind me of the advice my favorite clinical supervisor gave, because many of the stories we would hear as fledgling therapists were horrifying narratives of indifference, negligence, and cruelty. "Listen carefully," my mentor told me, "for the fragments in which a kindness or comfort was offered or experienced. These are the red threads of hope, tucked in us all as signs of resilience and the impulse to survive. It's with those fragile strands that healing can weave a new story." Stories of remorse, regret, contempt, and trauma can give rise to empathy, forgiveness, revelation, and redemption. Stories of love, compassion, and joy remind us of resilience, hope, and optimism. In storytelling, both light and shadow come into view, offering a fresh perspective on our lives and humanity.

Above all, stories can help us see that we are not alone in our struggles or triumphs. Conscious storytelling has the power to transport us from the personal to the universal. We crave the connection and personal growth it brings, which is evidenced in the proliferation of blogs, podcasts, poetry slams, and storytelling stages that capture the public imagination by meeting our collective hunger for authenticity and empathy. Stories bind us together, and the act of sharing can bridge the heart and mind, contribute to a sense of our own integrity, and ignite new possibilities. With the right conditions, a trusted listener, and a safe space, the power of story changes us for the good. It enables kindsight.

Writing Heals

It is well known that writing about difficult, shameful, and traumatic experiences is a powerful method for healing and personal transformation. Social psychologist James Pennebaker has studied the physical and emotional benefits of expressive writing for more than two decades. He has observed that writing about emotional life events in short durations sets off a cascade of positive effects, including better sleep, improved immune function, and less substance use. There is something about putting words to paper that can serve as a "course correction," because it gives structure and meaning to life's experiences, and it invites reflection. The fact that we can't write as fast as we think serves to slows us down—a profound effect in itself. It can be an incredibly kind thing to do for yourself. Here's some advice for how to write for yourself:

- **One week**: Pennebaker's approach is simple: Write for fifteen to twenty minutes daily, or every other day, for a week about stressful or traumatic experiences. He advises writing continuously, with no editing and no worries about spelling or grammar. You can tear it up afterward or revisit it later to polish it up.
- **Morning pages**: Julia Cameron, author of *The Artist's Way*, prescribes "morning pages" as a primary tool for creative recovery. This is a practice of producing three full pages of longhand, stream-of-consciousness writing upon first waking up in the morning, with no edits, corrections, or rewrites.
- **Rumbling**: Researcher and storyteller Brené Brown encourages "rumbling" with your story and not skipping the hard parts. "When it comes to the process of owning our hard stories, uncertainty can be so uncomfortable that we either walk away or race to the ending." While there may be discomfort, writing allows you to witness an experience as an observer and to process it in a fresh way.

When it comes to sharing what you write, Brown cautions to be careful: "We need to honor our struggle by sharing it with someone who has *earned* the right to hear it. When we're looking for compassion, it's about connecting with the *right person* at the *right time* about the *right issue*." Usually this doesn't mean having an audience and risking exposure but rather having one or two trusted confidants.

Kindness in Practice: *Journaling from the Heart*

Grab a notebook or a few pages of loose-leaf paper, plus a pen that is easy to write with, and put yourself in a comfortable atmosphere. Commit to writing for fifteen to twenty minutes without self-editing or making corrections. Set a timer so you have a clear start and finish, and plan to do something grounding afterward, such as taking a walk or cooking.

Reflect on a challenging life situation, past or present, that you would like to look at from a tender place of acceptance or understanding. These may be experiences that you have pushed aside, avoided, or taken for granted. Look at the situation with an open heart, and suspend any judgment. Even if there is no single event that comes to mind, reflect back on your life like a time traveler with curiosity. Here are lovely writing prompts adapted from some masters:

- **Reflect with compassion.** How is the experience you are reflecting on related to who you would like to become, who you have been in the past, or who you are now? How can you view this experience from a place of kindness or compassion?
- **Your future self.** Imagine yourself as a wise older version of yourself, eighty or ninety years old. Write a letter from your wise self to yourself at the age you are now. What do you tell your younger self about your inner strengths, courage, hope, and dreams? What encouragement and kindness do you offer? Or, if you prefer, write a letter from your eight-year-old self to you at your current age. What would you tell yourself from a child's mind of wonder?
- **Poetry.** Watch a poem unfold as you answer the prompt "I am from…" Reflect back on your life with openheartedness about places, people, culture, food, religion, beliefs, traditions, significant moments, and values. List them one after another. Be

descriptive about details such as colors, smells, sounds, or colloquial expressions. Notice any new or fresh connections that arise.

We live by the stories we have about our lives. We are storytellers, producers, moviemakers, songwriters, and journalists in one form or another. Usually, this occurs in the confines of our own mind. We are continually interpreting our words. Bringing stories to conscious awareness through writing and storytelling begins to loosen the hatch on some of the darker narratives that may lurk unattended.

Writing with intention and mindfulness serves to reinterpret your life experiences and can ignite profound change, connecting your head and your heart in new, expansive ways. It serves up many of the PEPPIE ingredients, especially *Presence, Perspective, Integration,* and *Purpose.* Just as Michael put words to paper about his evolution of kindness, so can you. It can turn hindsight into kindsight.

Reflection: *By writing down my stories, I can intentionally bring tender awareness to all my experiences, discover valuable lessons, and even rewrite the outcomes.*

CHAPTER 26

Accepting a Helping Hand

"I was ferociously independent, and I was frickin' proud of it," says Sam. "I never let myself be on the owing side of any favor. Somebody did me a nice, I made sure I did them two." Sam is an artist with the words "Make Make Make" tattooed across her forearm. Her self-reliant spirit was put to the test after she left a secure IT job to go back to school in her mid-forties to earn a master's degree in fine arts, a lifelong dream. When Sam finished, she was faced with a harsh reality: the worst economy in the United States since the Great Depression. There were no teaching jobs, and she was divorced, had a son off to college, and had a housing crisis on her hands. No job, no money, no house. As if that wasn't bad enough…

"I had a heart attack."

It was a shock like no other. Sam discovered she carried two genetic conditions that were fighting each other, leaving her heart vulnerable to failure. "You think everything's banging along just fine in there, and then *boom!* No, it's not." The medications that Sam takes to keep her heart condition in balance severely limit what she can do. "I started having to ask for help. I had to start saying, 'Can you please slow down? I can't walk that fast,' or 'I can't carry this. Can you help me?' or 'Doc, can you give me a handicap pass?' I had to start admitting to my fallibility and my humanity."

Then a friend who lived on the other side of the country said, "Just move in with me and we'll figure it out from here." Out of desperation, Sam showed up on her friend's doorstep with a moving crate. "She stuffed me in the tiniest bedroom in her house. Another friend, who had a large art studio close by, gave me a dinky room in his studio. That gave me a place to sleep and a place to be creative." In return, Sam agreed to help around her friend's house and make meals. They both thought it would be for three months. It lasted three years.

"She was kind. Kind beyond the call of duty. I came out of school alone, sick, and messy. I often call her my patron. My angel who let me start my quilting business, the angel who let me write my quilting book. The patron who let me figure it out." As the saying goes: *When life gives you scraps, make a quilt.*

A Friend to Count On

Social connection matters most in our darkest moments. An offer of help is a supreme gesture of compassion. And acceptance of help is a courageous act of self-kindness. It is a self-perpetuating circle of love and gratitude: Give. Receive. Appreciate. Give. Receive. Appreciate.

In a culture that idealizes rugged individualism rather than belonging to a community, we can be left feeling both alone and lonely. Yet, a shift happens when we realize we aren't separate from one another; rather, we are intimately connected. Sam's friend invited her into her arms and home with love. "There's this thing that I often say now: *You honor the people that love you by actually allowing them to do it.*" Sam's story shows how we are all stitched together, no matter what our life circumstances, conditioning, or culture. Like a beautiful quilt.

Novelist Anaïs Nin wrote, "All of my creating is an effort to weave a web of connection with the world. I am always weaving it because it was once broken." We need each other. Sometimes the simplest of "asks" can be hard to do, like asking for help carrying the groceries or accepting a room to live in. Sometimes the humblest and bravest thing is to accept help when it is offered. We can expand social connection and support when we are willing to be honest about our needs and compassionate with our suffering. That takes courage.

There is an important element in Sam's story that can't be denied: her gratitude. Sam was grateful for her "angel patrons" who helped her meet her basic need for shelter and food. She also deeply appreciated the opportunity to thrive, from the inside out, by being able to pursue her craft and creating a new livelihood. A kindfulness in the present moment and kindsight in retrospect make a mighty fine blend for future well-being.

Social Support Boosts Well-being

One person in your life, one social connection, can make a big difference. Social connection is broadly defined by social scientists as "a person's subjective sense of having close and positively experienced relationships with others in their social world." In effect, friendship and being in the company of others are good for health.

Its opposite is loneliness. Feelings of loneliness and isolation are associated with failing physical and emotional health. One in four Americans say that they have no one they can confide in about their trials or triumphs; and if you take family members out of the picture, the rate rises to half. It's unfortunate that these days, there are fewer ties to the neighborhoods and communities that once served as primary sources of social connection and exposure to people outside of family. Both *living alone* and *feeling lonely* are risk factors for mortality on par with the risks from smoking.

With social connectedness comes happiness and well-being. From an evolutionary point of view, we tend to seek friendships with people who are similar genetically and may boost the likelihood of being caring toward one another. Some of the active ingredients of social connection include having *a sense of similarity* with others, having an *emotional connection* that arises from empathy and helping behavior, and experiencing *affectionate social contact*. What matters most is your perception of the quality of a relationship or friendships, rather than the number of connections. The greatest benefits come from feelings of belonging and intimacy.

People with satisfying social relationships have higher levels of happiness, lower levels of depression and anxiety, and a higher likelihood of helping others. When friends share emotional experiences—positive or negative—they tend to lift each other up. Even our vocalizations—laughing, hootin', chortling, as well as crying and weeping—are contagious and evolutionarily adaptive for human bonding. Social connections serve as a buffer to stress by triggering the neural reward circuitry, and they ignite a physiological calming effect through regulating emotional distress and lowering stress hormones.

Expressing authentic concern for others and accepting help have a ripple effect that can improve everyone's health and happiness. And these actions create far greater possibilities for kindness than we can ever accomplish on our own.

Kindness in Practice: *Your Circle of Care*

The giving and receiving nature of kindness creates a positive feedback loop that you can initiate anytime. But sometimes it takes a little bit of effort to get that kindness cycle jump-started. It helps to get a view of the big picture.

> Map out your social connections on a piece of paper. At the top write "My Caring Network." In the center of the page write your

name. Create circles or nodes that represent people or communities of support in your life and label them, using the list of suggested categories here. Next, draw lines between circles indicating any connections starting with you and any branches between the nodes. Write the names of people you know in the circles. Some circles may be empty, and those areas point out opportunities to grow social connections with people and communities you trust.

Try to indicate at least one connection, even if you are unsure of the closeness of the connection. For instance, you might write in your doctor or pastor. Then consider to whom you might offer a helping hand or from whom you might ask for help when you are in need. Add as much detail as possible to your web of caring connections.

Me

Family

Friends Near

Friends Far

Neighbors

Coworkers

Helpers and Healers

Support Groups

House of Worship

Community Groups

Hobby

Interest Groups

People I Can Help

Volunteer Opportunities

The metaphor of a network of circles conveys containment, safety, sharing, and flow. You may surprise yourself with just how many names pop into your mind once you get going. When your circle is complete, set an intention to get in touch with a few of those people to either offer your help if you know they might be relieved to get it, or to ask for help if you think they're able to give it to you. Or just share your gratitude. It may seem unusual to reach out and give in this way, but everyone's likely to experience something positive as a result. "Giving is the great equalizer," writes Stephen Post, who has dedicated his life to researching the act of giving as a protective factor in health and well-being. "Whatever your background—privileged or impoverished, blessed or difficult—the starting place for a life of great love is within your reach."

There is nothing like a good friend to break a SPEL. Sometimes we are not even aware of who will be there until a calamity strikes. But why wait until that happens? Cultivating a deep friendship with one or two people is a surefire way to lift the spirits—and not just when life knocks you down. Creating a web of support is good for the heart and nourishes the soul.

As Sam realized, sometimes the kindest thing you can do is accept a helping hand, as surely there will be a time to offer a helping hand in return. We all need help, every single one of us. This truth is at the root of compassion. When you dig deep into your life history, tracing that red thread I described in chapter 25, you'll find that it's often the unexpected blemishes, imperfections, or mistakes that turn out to be part of the most beautiful embroidery of all. Because appreciating their place in the overall design helps us feel connected with each other, no matter what.

Reflection: *When I am at a loss for what to do next, I ask myself, "Who can I invite into my circle of care and trust? Who can I turn to for help?"*

CHAPTER 27

Gifts in Kind

Sometimes life doesn't turn out the way we thought it would. Rick explains that "Because of my family circumstances and the configuration of events that happened, I took a job at UPS thinking it would be temporary. It turned out to be a thirty-year career." Being a UPS deliveryman is hard, physically demanding work, especially around the winter holidays. But that's also the time of year that Rick looks forward to, because the tips helped him afford gifts for his own family. Typically, UPS hires part-time, seasonal help for the torrent of packages. One year, Rick's helper was Will, a shy nineteen-year-old who didn't say much.

Once Will got the hang of deliveries, Rick felt confident to one day leave him with four blocks of deliveries on the right side of the street. A half hour later, Rick met up with Will. "Will's face was beaming with a smile I'd never seen before. His packages were all gone, and as he got into the truck he was just glowing. So I asked how it went. He said, 'You know what? I got it done.' He was effervescent but didn't say anything, so I asked him to tell me what happened. Then Will said, 'Well, I went into the doctor's office and he gave me twenty-five dollars. I went into the dentist's office and there was a card there for me with fifty dollars. I went into the jewelry store and they handed me an envelope for twenty-five dollars!'"

It clearly didn't cross Will's mind that the tips, which in the end totaled $200, were meant for Rick. "I felt the extreme injustice of having him work for a week or two and taking these gifts that I should have gotten for a whole year of work. I was so close to pointing out to him that those tips were supposed to be for me." But Rick didn't. Maybe Will needed this money more and so he let it go, albeit a bit disgruntled. He never saw Will again.

"Three or four weeks later, I went into a bridal shop," Rick recalls. "The manager called me over and said he didn't get to see me for the holiday season, and handed me a card. I thanked him and went out to the truck. Inside was $200! I had never gotten so high a gift from one place, not even close. I was struck with the thought that my kindness had come right back to me, that a kind of reciprocal arrangement of the universe had occurred."

Rick may have been reluctant to give up his holiday tips, but he had a good heart. "I entered the job as a UPS man with way more education than was appropriate for what I was doing. I have two master's degrees, in divinity and in counseling. I tended to look at the neighborhoods on my route as my parishes, so to speak. These were my people, and I had a gift to bring, whether it was a package or just any kindness I could share." As he got to know customers, he was invited into their homes. He would eat meals with them or go to after-work events. "They became my friends, and I still keep in touch." It turns out Rick had been delivering happiness all along.

Delivering Kindness, One Act at a Time

Matthieu Ricard writes, "Loving-kindness and compassion are the two faces of altruism. It is their object that distinguishes them: loving-kindness wants all beings to experience happiness, while compassion focuses on eradicating their suffering. Both should last as long as there are beings and as long as they are suffering." Even though we aren't necessarily inclined to feel fondly or generous toward everyone we meet or hear about, we do have the ability to consider their struggles. We can also experience their joys. Ricard writes, "Altruistic love is characterized by unconditional kindness toward *all beings* and is apt to be expressed at any time in favor of *every being in particular.* It permeates the mind and is expressed appropriately, according to circumstances, to answer the needs of all."

Rick, while not having the faintest idea what Will's life was like, paused to consider this young man's needs and surrendered his own self-interests. This kind of altruism is the result of perceiving our common human needs. Ricard cites political scholar Kristin Renwick Monroe, who wrote, "Altruists simply have a different way of seeing things. When the rest of us see a stranger, altruists see a fellow human being." This is kind-sight that leads to acts of kindness. When you take care of your inner life with love and kindness, you become a beacon of light and a natural agent for change in the outer world. Service to others becomes natural. You plant the seeds to grow a kind world.

Kindness in Practice: *Pay It Forward, Catch It Backward*

The idea of "paying it forward" has gained traction in popular culture. Science confirms that friendliness, kindness, and generosity are incredibly good for you. Altruistic behavior makes people feel happier, triggering a "helper's high" in the brain; is good for love life; improves symptoms of chronic illness; alleviates social anxiety; can boost one's financial bottom line; and is associated with a longer life. In effect, paying it forward delivers the goodwill right back at ya.

Fortunately, there are many everyday kindness warriors paying a kindness forward by creating communities and broadcasting good deeds. In a global survey of empathy levels of 63 countries, including 100,463 adults, the United States ranked seventh (after Ecuador, Saudi Arabia, Peru,

Denmark, United Arab Emirates, and South Korea). The researchers found that "collectivist" countries (those with tighter kinship groups) were higher in empathy, and empathy across each country's volunteers was linked with agreeableness, conscientiousness, self-esteem, subjective well-being, and prosocial behaviors.

Wendi Gilbert, founder of *Take a Stand for Good*, has been tracking the social impact of kindness organizations, initiatives, school curricula, films, and individual kindness evangelists across the United States. She has identified more than seventy kindness-specific initiatives in the nation with a Facebook reach of more than 18 million followers, including the Pay It Forward Foundation, the Random Acts of Kindness Foundation, Think Kindness, and the World Kindness Movement, to name a few.

That doesn't include comedian Ellen DeGeneres, who has built her media platform on kindness and found her way into the hearts of millions of viewers by highlighting and supporting people's good deeds. In accepting a People's Choice Favorite Humanitarian Award, DeGeneres said, "I have to say, it's a little strange to actually get an award for being nice and generous and kind, which is what we're all supposed to do with one another. That's the point of being a human."

Jaclyn Lindsey, of kindness.org and kindlab, commissioned researchers at the Cognitive and Evolutionary Anthropology at Oxford University to review quality studies on kindness interventions, such as doing a certain number of kind acts over a week or spending money on someone else rather than on oneself. The review confirmed that when people perform acts of kindness at some cost to themselves, they are happier.

Researchers don't know precisely what characteristics or conditions motivate some people to be kinder than others. They just know that kind actions make people feel happier—and happier people do kind things. We could do more of them. As Tim McGraw's platinum hit song "Humble and Kind" conveys, there are many ways to do this. Lori McKenna, a mother of five and the first woman to win best songwriter in Academy of Country Awards history, wrote this song as a *reminder* for her children about what is important in life. It's fitting that a song with such a simple chord progression may be the language of kindness we needed to hear all along. It may be the song for our times.

Kindness in Practice: *Twenty Kindsight Questions*

Most people want to do the right thing, even when it feels hard. And it is at these very moments when our efforts really do matter. It's one thing to do the right or kind thing because it feels easy or automatic. But it is altogether a different thing when we need to rumble a bit on the inside, when our good deed comes at some cost to us—whether our emotions, time, or resources. Asking yourself questions will help you think about your capacity for kindness in new ways, like creating a list of reminders. You can also take the Kindness Quotient quiz on my website: http://www.taracousineau.com/whats-your-kindness-quotient/.

Here are twenty questions of the heart, the sort of questions that ask you to dig deeper and deeper until kindsight gets vaster. You can journal by yourself or spark a conversation with others on your next road trip.

1. What is one thing I can do today that will stretch my heart a bit wider?
2. What does a meaningful life mean to me?
3. If I knew I couldn't fail, what would I do?
4. What if my biggest fear came true? Then what?
5. What would I regret not doing at least once in my life?
6. What would I die for?
7. What am I most proud of?
8. How would my friends describe me?

9. Who is one person I can always count on?

10. If I could meet anyone in the world, who would it be?

11. What am I grateful for?

12. What is one habit I want to break, and what is one habit I want to create?

13. What does "god" or "spirit" mean to me?

14. How do I feel when I look up at a starry night?

15. What does nature mean to me?

16. What act or gesture of kindness did I do recently?

17. What are my biggest lessons in life?

18. What legacy do I want to create?

19. What does it mean to be enough?

20. When was the last time I said "I love you" to those I care about? To myself?

The fund-raiser and author Lynne Twist says, "What we appreciate, appreciates." Whether it is our time, money, or other resources, the spirit of generosity emerges when we realize that we are enough and that we have more than enough to give. Altruism becomes easier.

"I'm not the same man I was thirty years ago," concludes Rick. "I'm playing with the idea of *perfection*, that I am perfect and that everything that I am is perfect, and that my life has been perfect." In many aspects, Rick was indeed a master of divinity, cultivating his own brand of humble ministry from his decades of delivering goods and his kind service. But we are all imperfectly perfect underneath the layers of life's trials and triumphs—we just need to let our souls realize our potential for love and kindness.

When you get clear on values and recognize your inner worth, new possibilities abound. Kindsight renews faith in our ability to affect change. It is empowering. "When you let go of trying to get more of what you don't really need," writes Twist, "it frees up oceans of energy to make a difference with what you have. It expands." Generosity returns to you in ways you may not see right away, as the final chapter in this book shows.

Reflection: *My kindness reverberates in ways I am not aware of, and kindness may come back to me in ways I never expected.*

CHAPTER 28

Networks of Generosity

"One night when I was six years old," recalls Mann, "my siblings and I were awakened in the middle of the night. Our parents asked for us to be quiet and to get dressed quickly. They told us not to bring any of our belongings with us since we wouldn't be coming back to our home." Mann's extended family was among a mass exodus of people fleeing Vietnam in 1976. Her story of survival is the result of a helping hand at every turn.

After two years in a Malaysian refugee camp, the family was selected by lottery for passage to the United States. Her mother, eight months pregnant and severely malnourished, snuck through health inspections so that they wouldn't be detained. Initially, Mann's family lived with a sponsoring family until they could access government assistance and food stamps. Then another Vietnamese family, who had been in the same position eight years earlier, gave them money to live. "These people didn't know my parents and gave them $5,000, which was a significant amount of money at that time. When my parents eventually earned the money, they offered it back to the couple. They said, 'No, just pay it forward. Give it to someone else.'"

The generosity kept coming. Mann's baby sister was born with a hole in her heart and needed medical care they could not afford. "We were introduced to a couple, Judith and Edward. Edward was the head nurse in an emergency room. We were fortunate to meet them, because they helped us get the care needed for my sister. She had her open-heart surgery for free. Then they let us live on their first floor, in a two-bedroom apartment, also for free. In return, my parents would help them clean. Then they found positions for my mom to care for other young children in the neighborhood. These were high-net-worth individuals who treated my family like their own." Remarkably, these local families offered to pay for Mann and her sisters' private schooling, but her parents were too proud. However, her parents did allow the girls to do one thing these families offered. "The other families sent their children to finishing school, and for giggles my parents thought, 'Okay, why don't you go learn to be more well mannered?'"

Mann's family worked hard and thrived. Eventually, her extended family owned restaurants and supermarkets in Chinatown. "We believe in 'you do good' regardless of how nasty or unkind someone is to you—you still move forward in being kind and generous." Her family helped other refugees, allowing them to stay with them until they also found housing and work. "There are three things that are important in my family: love and kindness, filial piety—which is respect for others—and being generous and giving. If you have five dollars, give half. If you have something to eat, give half. Because you don't know when you might be in a situation when you need someone to be kind to you in return."

Coming to the United States was a long and hard journey for Mann's family. She says, "I've had so many opportunities, coming as a little girl and learning English here. Every year, every moment, everything that I do in my life, I feel like I've always had very positive support growing up. I think that if you give, you will receive back. I don't intentionally give to want to receive; I give because I want to give."

A Common Humanity

"We are all part of humanity, and each of us has the responsibility to improve humanity and to bring it additional happiness in order to make it more peaceful, friendlier, and compassionate," said the Dalai Lama. The story of the Vietnam exodus is both daunting and astounding. Through the goodwill of other nations, many refugees in Southeast Asia resettled across the globe. The global community became good neighbors. But even that changes from generation to generation. Right now, in various corners of the world, millions of people are fleeing their homelands or are displaced, searching for safety, food, and shelter.

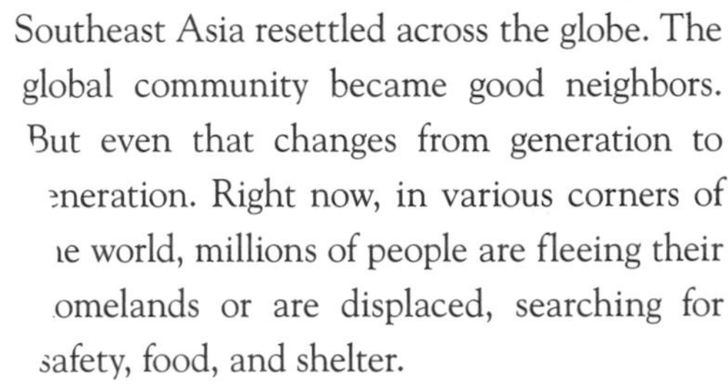

When I hear these refugee stories, I think of Mann's family and the various connections: from a fishing boat that survived the high seas to a refugee camp to a neighborhood in Boston—all with support from strangers. There seemed to be a series of "miracles," yet what was at work was an intricate chain of human goodwill.

Cooperation Is Contagious

We tend to copy the behaviors of people we care about—whether they are good or bad. And it's more than mere imitation. We also spread expectations and beliefs about appropriate behaviors—what scientists call *social norms*. The study of human social networks—whether they are face-to-face or online—shows that behaviors, ideas, and emotions spread among people with social ties.

In their book *Connected*, Nicholas Christakis and James Fowler explain how they mathematically connect the dots between complex human relationships, and they show how a person's mood, health status, or action spreads. In effect, they can track what it means to pay it forward, whatever the "it" may be. For example, if a person smokes, is overweight, gets divorced, makes a donation, or goes to the polls to vote, these very attitudes or behaviors will influence friends to act similarly. There is a multiplier effect. The researchers have demonstrated this in a laboratory using "public goods" games with strangers. If the first person cooperates (gives money away to other people in future games whom they have not met), this generosity spreads to three other people, who each then spread it to three more, and so on. They call this phenomenon the Three Degrees of Influence Rule.

This shows that if you demonstrate a kindness even when it is at a cost to you, that generous behavior spreads to your friend (one degree), your friend's friend (two degrees), and your friend's friend's friend (three degrees)—reaching people you don't even know. Similarly, that third-degree friend you don't know can influence you too, just by being in a network of shared social contacts. Christakis suggests that we assemble ourselves as "super organisms," meaning we are organically connected to one another with emotions, beliefs, and memories. Our networks, he believes, are a kind of social capital.

The upside is that acts of kindness, generosity, and cooperation can spread with only a few people. Of course, the opposite can also happen: networks can spread harmful ideologies and behaviors, like fascism or terrorism. It's why I felt so overcome by looking at the German children's books at the Holocaust Memorial Museum that inspired this book. Of course, this is the very reason why Mann's family sought refuge in the first place, and why millions of others flee from war and violence every single day.

"Social networks magnify whatever they are seeded with." Ultimately, Christakis is hopeful:

> We form social networks because the benefits of a connected life outweigh the costs. If I was always violent toward you or gave you misinformation or made you sad or infected you with deadly germs, you would cut the ties to me and the network would disintegrate. So the spread of good and valuable things is required to

> sustain and nourish social networks. Similarly, social networks are required for the spread of good and valuable things like love and kindness and happiness and altruism and ideas.

Kindness in Practice: *Acts of Generosity*

Knowing that what you do can influence at least three other people, what attitude or act of kindness, goodwill, or social action do you want to propagate? Sometimes, kindness toward yourself is the best way to begin the chain of kindness. What's the one thing you can do today? Here are some ideas, both big and small, to get you started. If you want more ideas, you can check out kind-minded organizations in the Resources section of my website: http://www.taracousineau.com/resources.

Kindness Toward Self	*Kindness Toward Others*
Smile at yourself.	Smile at others.
Say "I love you" to yourself.	Say "I love you," offer a hug, or give a high five.
Write a positive message about your body on a sticky note and post it on your mirror.	Send a thank-you note or e-mail to one person a week.
Eat healthy food and be physically active.	Open the door for someone, buy a cup of coffee, or pay a toll.
Find time to rest, play, and be creative just for fun.	Compliment three people for their positive attitude or efforts.
Write a letter to your future self about hopes and dreams.	Get to know someone who is not like you.

Start a gratitude journal.	Set a daily kindness reminder on your phone.
Keep your home neat and tidy.	Donate to a cause and let others know you did (kindness is contagious).
Sing with others.	Sign a social action petition or do something to impact social change.
Download and listen to a self-compassion meditation or guided imagery.	Volunteer at a soup kitchen or food bank.
Post a daily or weekly positive affirmation on social media.	Enlist friends to do something nice for someone.
Listen to inspirational and diverse stories and podcasts such as *Kind World by WBUR* , *The Moth Radio Hour,* or *StoryCorps*.	Pick up trash on your street and in your neighborhood.
Watch *The Ellen Show*	Watch *The Ellen Show* with friends.

It's encouraging, and downright inspiring, to realize that an act of caring and generosity reverberates in ways you may never see. Knowing the fact that your kindness is contagious turns kindness into personal responsibility for the greater good. It moves you up into the range of high empathy and high empowerment, warding off any old SPEL that blocks your way.

"Now that I have all this stuff in my life," reflects Mann, who is starting a food pantry where she lives, "and all the things of privilege, I always refer back to the small moments in my life that were the happiest, when I had nothing. Everything seemed like a gift back then. There was one moment when us girls were with my baby sister listening to Michael Jackson's hit song 'Rock with You,' huddled around a brown radio box. We were so excited to be in America, to be safe. To sing in English! That was what mattered to me in life."

Kindness is simple and sacred. Sometimes the most generous thing we can do is acknowledge the struggles, appreciate the simple gifts, and choose to give back—just like Mann and her sisters, sharing the beat of love.

Reflection: *Kindness is contagious. No matter how my day unfolds I will carry this thought: "Be kinder than you think you are" and see what happens.*

CONCLUSION

Reimagining Kindness

Your kindness instinct runs deep and wide. I've cast a wide net to see how kindness serves as a cure. You can ignite kindness in so many ways, through practices of kindfulness and kindsight, compassion, gratitude, forgiveness, and generosity. Kindness starts from the inside out through daily self-care practices, and it grows in the presence of caring people and caring communities. Along the way, you've considered:

What triggers my kindness instinct?

What conditions rekindle it and ignite my caring circuitry?

How can I bring kindfulness into my life through self-compassion and loving-awareness?

Can I allow forgiveness to release me from suffering, and gratitude to connect me to joy?

How can I bring kindfulness into the present moment?

What life stories can I rewrite from a lens of kindsight?

Who is in my circle of care?

This inquiry also begs harder questions at the heart of the matter: What gets in the way of kindness? Who am I leaving out? I often reflect on the writer George Saunders's poignant advice to college graduates at Syracuse University:

> What I regret most in my life are *failures of kindness*. Those moments when another human being was there, in front of me, suffering, and I responded...sensibly. Reservedly. Mildly.
>
> Or, to look at it from the other end of the telescope: Who, in *your* life, do you remember most fondly, with the most undeniable feelings of warmth? Those who were kindest to you, I bet. It's a little facile, maybe, and certainly hard to implement, but I'd say, as a goal in life, you could do worse than: *Try to be kinder.*

Easier said than done. Kindness can get a bad rap. Kindness is not a sugary coating we can slather on everything and everyone in our life. It's true that kindness can feel hard to engage, it can challenge us, and it can hit a multitude of emotional triggers in ourselves and others. When it does, we can feel like kindness falls flat or even fails us. Yet at the same time, people are crying out for its resurrection. That's the real kindness crux.

A high school friend, Patty, who lives in Newtown, Connecticut, poignantly observed, "After the shooting, our community took on the 'be kind' mantra. Sandy Hook's principal's call to action was, 'Be kind to one another; it's all that really matters.' There were shirts, bumper stickers, signs in shop windows; random gift cards or homemade treats left on car windshields or brought to police and fire departments; doors being opened; and smiles and hellos to everyone you passed. So many wonderful acts of kindness were displayed and are still happening today. It makes me sad that it took a horrific tragedy for this to happen."

A Kind Orientation to Life

When tragedy strikes, all we can do is offer kindness. But when the little, daily uncertainties and fears arise, it can feel too hard to be kind and compassionate. Likewise, when our best intentions and kind actions are met with cruelty and contempt, we can feel defeated. It is precisely in these moments when training in kindfulness is the call to action. This requires *reimagining our relationship to kindness.* It's orienting yourself to a way of being that is caring, compassionate, and connected—and ultimately liberating.

The basis for this book has been the view that kindness is *strong,* not weak. It is *courageous,* not cowardly. It is *heartful,* not heartless. It's not about being nice, agreeable, or virtuous. It's about understanding, having boundaries, and taking reasonable action. It's moving from empathic distress to motivational empathy. Taking a kind stance doesn't mean giving in.

A protest slogan reads: "Feel the Rage, Be the Love." These six small words perfectly capture both the challenge and the solution when facing difficult persons, places, or things. The meditation teacher Sharon Salzberg asks, "Why can't we both love and resist at the same time?" This is a perfect question for our life and our times, and it forms the basis for reimagining kindness.

The Prescription

The *Kindness Cure* is not just for you—it's for all of us—the super-caregiving species that we are designed to be. Without even being aware of it, we can be caught under a SPEL: *Self-Protective Empathy Lethargy.* A SPEL keeps us from the potential of a more expansive reality where we can both *believe* and *dwell* in a kind and loving world. It keeps us from the full potential of happiness and well-being.

The antidote to a SPEL is the mix of ingredients that serves to transform us from surviving into thriving, to finding that sweet spot between empathy and empowerment. The prescription is PEPPIE, and it's up to you to decide how you want to practice it.

P*resence*

E*motional regulation*

P*erspective*

P*urpose*

I*ntegration*

E*ffort*

I hope you have tried some of the "Kindness in Practice" exercises in this book. In order to strengthen your kindness instinct, you need to strengthen those neural pathways—through daily practices and regular doses of kindness, caring, and compassion. While you can't save the world with kindness, you can begin with who you know and what pulls on your heartstrings. After all, kindness radiates in ways you may never see.

We need to reimagine kindness as both a virtue and a way of being. The spiritual teacher Marianne Williamson said, "The way of the miracle worker is to see all behaviors as one of two things: either love or a call for love." Can you reimagine *kindness as love in action*?

The Kindness Manifesto is a call to put love into action. You can download a copy at http://www.taracousineau.com/manifesto.

Kindness Manifesto

You are wired to care.

Humans have evolved not just to survive but to thrive. We are born with a unique physiology for caring, compassion, and cooperation. Kindness lives deep in our bones and in our DNA. You have an instinct for compassion.

We can grow kind minds.

Our hearts are wise and our brains are flexible. By comforting others and being comforted, you can strengthen your caring neural circuitry. You can kindle and rekindle kindness.

Don't let kindness become a dwindling inner resource.

We can get caught in a spell that separates us from the rest of humanity: **S**elf-**P**rotective **E**mpathy **L**ethargy: SPEL. It's sneaky. Without being aware, anyone can be snared in a culture of indifference, a mindset of scarcity, and a fear of others. Golden rules start to fade. Or else we feel too much. Sometimes we believe that there's so much suffering that we don't know where to begin, how to help, or how to make a difference. We turn away. We risk losing our innate connection to love and kindness.

Together, we can grow kindness.

We can awaken to compassion. We can raise our levels of empathy and empowerment for our own good, for the greater good. There's a balance. A sweet spot. You can break the trance with an elixir of ingredients:

P*resence,* **E***motional regulation,* **P***erspective,* **P***urpose,* **I***ntegration,* **E***ffort.*

Get PEPPIE. Be enchanted with kindness.

Shift from stressed to blessed.

Be kind to yourself.

It starts with *you*. Breathe. Meditate. Relax more. Linger in the moment. Learn to love silence. Spend time in nature. Practice kindfulness. Expand kindsight. Let the energy of tenderness heal you. Radiate loving-kindness.

Choose to believe that the world is a friendly place.

Small acts of kindness begin at home. Say: "I love you," "Please," "Thank you," "I'm sorry." Listen with compassion. Know your boundaries. Trust. Have faith. Connect to core values. Be gentle. Touch. Cherish the little things. Practice gratitude. Forgive. Give. LOVE. Play nice. Sing. Dance. When you become kinder to yourself, you find compassion for others and for our planet.

Be kinder than you think you are.

If you find the world lacking in kindness, then you have work to do. Ask how you can help. Be a mentor. Be a role model. Walk the talk. Step into another's shoes. Put in joyful effort. Increase your kindness quotient: KQ.

Kindness is contagious.

Kindness matters. You may never see the far-reaching ripples of your kindness, but toss the pebble anyway. Be outrageously kind even if you are met with suspicion. Love anyway. Pay it forward. Catch it backward. Build kind communities. Make kindness cool. Find the kind and courageous in history and emulate that. When people are kind-hearted, generosity spreads and people are happy. Scientists say so. Ancient wisdom has always known it.

Kindness is love in action.

Maybe your act of kindness won't win the war on despair and indifference, but you can train a few soldiers in the process. Focus on what has beauty and dignity among humankind. Lead with a kind and brave heart.

Be a kindness warrior.

To be human is to be kind.

You have all the gifts inside to spark a kindness cure. Follow the KISS principle:

K*indness* **I***s* **S***imple and* **S***acred.*

Acknowledgments

I am so grateful to all the people who opened their hearts to me, whether over many years or over a brief exchange, and shared their *kindsights*: stories of struggle and triumph. You are true kindness warriors.

It's been a privilege to learn from master teachers in psychology, mind-body medicine, and the healing arts, whether through supervision, professional trainings, workshops, or private sessions. I am especially thankful to Rick Hanson, who taught a positive neuroplasticity workshop one Cape Cod summer and advised me to write my book one blurb, one page, at a time, strengthening those neural networks for writing. Good advice for a novice. I'm grateful for the mindfulness and spiritual teachings of Tara Brach, Jon Kabat-Zinn, Jack Kornfield, Matthieu Ricard, and Dan Siegel; and mentors at the Center for Mindfulness and Compassion in Cambridge, Massachusetts, including Zev Schuman-Olivier, Christopher Germer, Susan Pollak, and Christopher Willard; and my esteemed colleagues who have supported me throughout the years: Bornali Basu, Joe Burton, Alice Domar, Liz Donovan, Nancy Etcoff, and Debbie Franko.

I am deeply grateful to my spiritual support team: Janet Conner, Nicky Robertson, Reiki Master Patricia Iyer, Margo Mastromarchi, and Jennie Mulqueen; and my Woman Song sisters at the South Shore Conservatory in Hingham, Massachusetts. Finding my voice has been a lifelong process, and I am indebted to K.C. Baker and Marsha Shandur for helping me speak my truth. I could never have endured the many hours of writing if it were not for the Universal Power Yoga community and the teachers who truly understand the power of breath, mindfulness, and movement to ignite our innate compassion circuitry: Jen Palmer, Sue Bonanno, Jacqui Bonwell, Elyse Callahan, Hania Khuri-Trapper, and Jamie and Bill MacDonald.

I could not have done this without masterful editing by Jennifer Holder, whose empathic guidance channeled *kindfulness* onto the page, and the wonderful editors and designers at New Harbinger, including Marisa Solís; Clancy Drake; Vicraj Gill; Amy Shoup; and acquisitions editor Jess O'Brien, who believed in this project. Thanks to the careful readers who helped with early versions of the manuscript, including Tom Guilmette, PhD, and Caroline Pincus; a deep gratitude to my literary agent, JoAnn Deck, for seeing me through; and for the patient assistance of Amy Sosa. Illustrator Pamela Best, who is an aspiring neurobiologist, deserves special recognition for her drawings and bringing this book's uplifting messages to life.

Gratitude is due for the funky little moments with friends who have had profound reverberations in my life and prove that kindness is contagious: Mary Dickie for inviting me to meet Matthieu Ricard; Kristen Darcy, Elaine Freibott, Christine Egan, Celine McDonald, Heidi Melsheimer, Kate Sweetman, and Renée Wilson; and my "allomoms" Mrs. Mel and Mrs. Katis.

My mother, a former Avon lady, was my first teacher of compassion. With my sister, Tina, and I in tow she demonstrated and supported the beauty and dignity among humankind in the small towns, door to door, across the Connecticut shore. Much love and appreciation go to Tina, my niece and nephews, and the Cousineau clan for their support, kindness, and good cheer over many years.

A thank-you to my beloved husband, Steve Cousineau, whose unconditional love and patience have seen me through more than half of my life, and who makes good on his personal mission to make at least one person laugh every day. And finally, to our daughters, Sophie and Josie, I wrote this for you.

Resources

Please visit my website, http://www.taracousineau.com, and blog for continued stories and news about kindness, resources, articles, and updates to the Kindness Manifesto, which serves as a living document.

Notes

Introduction

3. **"kindness is always hazardous":** Adam Phillips and Barbara Taylor, *On Kindness* (New York: Picador, 2009), 5.
4. **"people are leading secretly kind lives":** Phillips and Taylor, *On Kindness*, 4.
5. **"when children do not prioritize caring:** Rick Weissbourd, PhD, personal interview with author, August 10, 2016; Making Caring Common, *The Children We Mean to Raise: The Real Messages Adults Are Sending about Values* (2014), http://mcc.gse.harvard.edu/files/gse-mcc/files/mcc-research-report.pdf.

Chapter 1: Kindness Takes Effort

10. **kindness as a mode of altruism:** Matthieu Ricard, *Altruism: The Power of Compassion to Change Yourself and the World* (New York: Little, Brown and Company, 2015).
11. *mettā:* Sharon Salzberg, *Loving-Kindness: The Revolutionary Art of Happiness* (Boston: Shambhala Publications, 1995), 21.
12. **positive emotions broaden and build your inner resources:** Barbara L. Fredrickson, *Positivity: Groundbreaking Research Reveals How to Embrace the Hidden Strength of Positive Emotions, Overcome Negativity, and Thrive* (New York: Crown Publishers, 2009); Barbara L. Fredrickson, "Positive Emotions Broaden and Build," in P. Devine and A. Plant (eds.) *Advances in Experimental Social Psychology,* volume 7 (San Diego, CA: Academic Press, Elsevier, 2013), 1–53.
14. **empathic distress:** O. Klimecki and T. Singer, "Empathic Distress Fatigue Rather Than Compassion Fatigue? Integrating Findings from Empathy Research in Psychology and Social Neuroscience," in *Pathological Altruism,*

ed. B. Oakley, A. Knafo, G. Madhavan, and D. S. Wilson (New York: Oxford University Press, 2011), 368–83.

Chapter 2: Your Kindness Instinct

21. **a warrior's mind:** Chögyam Trungpa Rinpoche, as cited in *The Essential Mystics, Poets, Saints, and Sages: A Wisdom Treasury*, ed. R. Hooper (Charlottesville, VA: Hampton Roads Publishing Company, 2013), 182.

22. **sympathy is a reflexive social instinct:** Dacher Keltner, *Born to Be Good: The Science of a Meaningful Life* (New York: W. W. Norton & Company, 2009).

22. **compassion and kindness "embedded into the folds of our brains":** Dacher Keltner, "The Compassion Instinct," in *The Compassion Instinct: The Science of Human Goodness*, eds. D. Keltner, J. Marsh, and J. Smith (New York: W. W. Norton & Company, 2010), 10.

22. **"innate, basic goodness":** Richard J. Davidson, "The Four Keys to Well-Being," Greater Good Magazine, March 21, 2016, http://greatergood .berkeley.edu/article/item/the_four_keys_to_well_being.

Chapter 3: Happy to Help

25. **Dylan's story:** Erica Lantz, "So Chocolate Bar," episode #22, NPR's *Kind World*, April 14, 2016, http://www.wbur.org/kindworld/2016/04/14 /kind-world-22-so-chocolate-bar.

26. **we are wired for empathy:** Mary Gordon, *The Roots of Empathy: Changing the World Child by Child* (New York: The Experiment, LLC, 2009).

26. **two basic functions in the mind:** T. Singer, "The Neuronal Basis and Ontogeny of Empathy and Mind Reading: Review of Literature and Implications for Future Research," *Neuroscience and Biobehavioral Reviews* 30 (2006): 855–863.

27. **toddlers...show kind helping behaviors:** K. A. Dunfield and V. A Kuhlmeier, "Classifying Prosocial Behavior: children's responses to instrumental need, emotional distress, and material desire." *Child Development* 84 (2013): 1766–76; M. Svetlova, S. R. Nichols, and C. A. Brownell, "Toddlers' Prosocial Behavior: From Instrumental to Empathic to Altruistic Helping," *Child Development* 81 (2010): 1814–27; C. Zahn-Waxler and M. Radke-Yarrow. "The Development of Altruism: Alternative Research Strategies." In N. Eisenberg (Ed.), *The Development of Prosocial Behavior* (New York: Academic, 1982), 109–37.109–137.

27. **children readily enjoy helping others:** F. Warneken and M. Tomasello, "Helping and Cooperation at 14 Months of Age," *Infancy* 11 (3, 2007): 271–294.

27. **toddlers' spontaneous helping behaviors:** Felix Warneken, PhD, "The Roots of Altruism," presentation given at the Dalai Lama Center, Vancouver, Canada (2014) http://dalailamacenter.org/heart-mind-2014-science-kindness/heart-mind-2014-presentations/felix-warneken.

28. **labeling emotions:** N. Moyal, A. Henik, and G. E. Anholt, "Cognitive Strategies to Regulate Emotions—Current Evidence and Future Directions," *Frontiers in Psychology* 4 (2013): 1019; M. D. Lieberman et al., "Putting Feelings into Words: Affect Labeling Disrupts Amygdala Activity in Response to Affective Stimuli," *Psychological Science* 18 (5, 2007): 421–8. http://www.scn.ucla.edu/pdf/AL(2007).pdf

29. **feelings word list:** G. E. Joseph and P. S. Strain, "Enhancing Emotional Vocabulary in Young Children," *Young Exceptional Children* 6 (4, Summer 2003): 18–26, http://csefel.vanderbilt.edu/modules/module2/handout6.pdf.

Chapter 4: Cultivating Courage

32. **vulnerability:** Brené Brown, *Daring Greatly: How the Courage to Be Vulnerable Transforms the Way We Live, Love, Parent, and Lead* (New York: Gotham Books, 2012), 45.

33. **little kind act:** Erika Lantz, producer, *Kind World*, WBUR, Boston, personal interview with author, March 3, 2017.

33. **courage to be kind:** Rick Weissbourd, PhD, personal interview with author, August 10, 2016.

34. **instinct to protect ourselves:** Parker J. Palmer, *A Hidden Wholeness: The Journey Toward an Undivided Life* (San Francisco: Jossey-Bass, 2004), 14–15.

36. **"our common human hospitality":** Greg Boyle, *Tattoos on the Heart: The Power of Boundless Compassion* (New York: Free Press, 2010), xv.

Chapter 5: Compassionate on Purpose

38. **altruism defined:** N. Moyal, A. Henik, and G. E. Anholt, "Cognitive Strategies to Regulate Emotions—Current Evidence and Future Directions," *Frontiers in Psychology* 4 (2013): 1019.

39. **"the joy of giving" has an anatomical basis in the brain:** M. D. Lieberman et al., "Putting Feelings into Words."

39. **five elements of compassion:** C. Strauss et al., "What Is Compassion and How Can We Measure It? A Review of Definitions and Measures," *Clinical Psychology Review* 47 (2016): 15–27, http://dx.doi.org/10.1016/j.cpr.2016.05.004.

40. **psychologists use loving-kindness meditation to develop wellness skills:** X. Zeng et al., "The Effect of Loving-Kindness Meditation on Positive Emotions: A Meta-Analytic Review," *Frontiers in Psychology* 6 (2015): 1693.

Chapter 6: Imagination's Objects of Affection

44. **"Sewing Hope":** YouTube video, 5:45, posted by David Pires, October 26, 2016, https://www.youtube.com/watch?v=n1Pbu2ZANGk; R. Chaleyer, "'He Looks at Sadness and Turns It Upside Down': 12-Yr-Old Campbell Makes Teddy Bears for Sick Kids," *The Feed*, October 24, 2016, http://www.sbs.com.au/news/thefeed/article/2016/10/24/he-looks-sadness-and-turns-it-upside-down-12-yr-old-campbell-makes-teddy-bears.

45. **applied imagination:** Sir Ken Robinson, "Imagination and Empathy," presentation at the Dalai Lama Center Speaker Series *Educating the Heart and Mind*, Dalai Lama Center for Peace and Education, November 21, 2011, https://www.youtube.com/watch?v=Yu2zcmb3yAQ.

46. **transitional objects:** D. W. Winnicott, "Transitional Objects and Transitional Phenomena: A Study of the First Not-Me Possession," *International Journal of Psycho-Analysis*, 34 (1953): 89–97.

46. **fantasy as an aspect of empathy:** M. H. Davis, "A Multidimensional Approach to Individual Differences in Empathy," *JSAS Catalog of Selected Documents in Psychology* 10 (1980): 85 http://www.uv.es/~friasnav/Davis_1980.pdf; M. H. Davis, "Measuring Individual Differences in Empathy: Evidence for a Multidimensional Approach," *Journal of Personality and Social Psychology* 44 (1983): 113–126; M. Taylor, *Imaginary Companions and the Children Who Create Them* (New York: Oxford University Press, 2001); T. Gleason and L. Hohmann, "Concepts of Real and Imaginary Friendships in Early Childhood," *Social Development* 15 (2006): 128–144.

46. **imaginary friends and empathy:** Tracy Gleason, PhD, Associate Professor of Psychology, Wellesley College Child Study Center, Wellesley, MA, personal interview with the author, December 21, 2016.

47. **parasocial relationships:** Tracy Gleason, personal interview with the author, December 21, 2016.

47. **a bereaved widow will commonly imagine:** S. M. Dannebaum and R. T. Kinnier, "Imaginal Relationships with the Dead: Applications for

Psychotherapy," *Journal of Humanistic Psychology* 49 (1, 2009): 100–113; N. P. Field, E. Gal-Oz, and G. A. Bonanno, "Continuing Bonds and Adjustment at 5 Years After the Death of a Spouse," *Journal of Consulting and Clinical Psychology* 71 (1, 2003): 110–117.

48. **recovering from disappointments and traumas:** J. P. Wilson and J. D. Lindy, *Trauma, Culture, and Metaphor: Pathways of Transformation and Integration* (New York: Routledge, 2013), 12.

Chapter 7: The Kinship of Belonging to Each Other

50. **Treehouse Community:** Judy Cockerton, founder of the Treehouse Foundation, Re-Envision Foster Care Initiative, personal interview with the author, December 16, 2016; Brian MacQuarrie, "In Easthampton Village, Everyone Helps the Children," *Boston Globe*, December 21, 2015, https://www.bostonglobe.com/metro/2015/12/20/village-raise-adopted -foster-children/GoqaPeIrxsZieXd39YMPkM/story.html.

52. **"intimate mutual belonging":** Joanna Macy, "Intimate, Mutual Belonging," Global Oneness Project video, 2:24, https://www.globalone nessproject.org/library/interviews/intimate-mutual-belonging and https:// www.globalonenessproject.org/people/joanna-macy.

52. **caregiving genes and allomothering:** Sarah Blaffer Hrdy, *Mothers and Others: The Evolutionary Origins of Mutual Understanding* (Cambridge, MA: Harvard University Press, 2011).

53. **impact of stressful childhood experiences on physical and mental health:** J. L. Hanson et al., "Behavioral Problems After Early Life Stress: Contributions of the Hippocampus and Amygdala," *Biological Psychiatry* 77 (2015): 314–323; C. E. Hostinar et al., "Frontal Brain Asymmetry, Childhood Maltreatment, and Low-Grade Inflammation at Midlife," *Psychoneuroendocrinology* 75 (2017): 152–163.

53. **Adverse Childhood Experiences (ACE) Study:** V. J. Felitti, R. F. Anda, and D. Nordenberg, "Relationship of Childhood Abuse and Household Dysfunction to Many of the Leading Causes of Death in Adults," *American Journal of Preventive Medicine* 14 (1998): 245–258.

53. **interpreting ACE scores:** Nadine Burke Harris, MD, "How Childhood Trauma Affects Health Across a Lifetime," TEDMED video, 15:59, September 2014, https://www.ted.com/talks/nadine_burke_harris_how _childhood_trauma_affects_health_across_a_lifetime; Robert Anda, MD, in *Resilience: Biology of Stress and Science of Hope* (United States: KPJR Films, 2016), James Redford (director and producer), Karen Pritzker (writer and producer).

53. **ACE questionnaire:** Aces Too High, "Got Your ACE Score?" https://acestoohigh.com/got-your-ace-score.

54. **compassionate responses to ACE scores:** L. C. Vettese, C. E. Dyer, W. L. Li, and C. Wekele, "Does Self-Compassion Mitigate the Association Between Childhood Maltreatment and Later Emotional Regulation Difficulties? A Preliminary Investigation," *International Journal of Mental Health Addiction* 9 (2011): 480–91; A. Hoffart, T. Øktedalen, and T. F. Langkaas, "Self-Compassion Influences PTSD Symptoms in the Process of Change in Trauma-Focused Cognitive-Behavioral Therapies: A Study of Within-Person Processes," *Frontiers in Psychology* 6 (2015): 1273.

Chapter 8: Reset Your Stress

60. **Justin Blaze and VEToga:** Justin "Blaze" Blazejeweski, founder of VEToga, personal interview with founder, September 29, 2016; "Justin of VEToga Talks Kindness," Lululemon video, 2:05, November 15, 2015, http://vetoga.org/video-justin-of-vetoga-talks-kindness; "VEToga Teacher Training," YouTube video, 2:40, posted by VEToga Organization, January 26, 2017, https://www.youtube.com/watch?v=Ko2Bitc3DII.

61. **survival instinct is found in the limbic system:** H. J. Markowitsch and A. Staniloiu, "Amygdala in Action: Relaying Biological and Social Significance to Autobiographical Memory," *Neuropsychologia* 49 (2011): 718–33.

61. **high levels of stress hormones have serious physical and mental health repercussions:** R. Sapolsky, "Stress and Plasticity in the Limbic System," *Neurochemical Research* 28 (2003): 1735–42; R. M. Sapolsky, "How to Relieve Stress," Greater Good Science Center, March 22, 2012, http://greatergood.berkeley.edu/article/item/how_to_relieve_stress.

62. **"stress levels are at an unprecedented high":** American Psychological Association, *Stress in America: Coping with Change*, Stress in America Survey, 10th ed., February 15, 2017, http://www.apa.org/news/press/releases/stress/2016/coping-with-change.PDF.

62. **autonomic nervous system:** Dan Siegel, "Brain Insights and Well-Being," DrDanSiegel.com, January 22, 2015.

63. **"tend and befriend":** Shelley E. Taylor et al., "Biobehavioral Responses to Stress in Females: Tend-and-Befriend, Not Fight-or-Flight," *Psychological Review* 107 (2000): 411–29.

65. **yoga and mindfulness…beneficial for veterans with PTSD:** K. Dahm et al., "Mindfulness, Self-Compassion, Posttraumatic Stress Disorder

Symptoms, and Functional Disability in U.S. Iraq and Afghanistan War Veterans," *Journal of Traumatic Stress* 28 (2015): 460–464.

Chapter 9: Befriending Your Senses

67. **"the body is the house in which the spirit resides":** Rumi, "The Body Is the House in Which the Spirit Resides," in *The Essential Mystics, Poets, Saints, and Sages*, ed. R. Hooper (Charlottesville, VA: Hampton Roads Publishing, 2013), 240.

67. **"presence is the awareness that is intrinsic to our nature.":** Tara Brach, *True Refuge: Finding Peace and Freedom in Your Own Awakened Heart* (New York: Bantam Books, 2016), 12.

68. **interoception:** Steven W. Porges, *The Polyvagal Theory: Neurophysiological Foundations of Emotion, Attachment, Communication, and Self-Regulation* (New York: W. W. Norton & Company, 2011), 75–82.

69. **"a fear of fear itself":** Bessel van der Kolk, *The Body Keeps the Score: Brain, Mind, and Body in the Healing of Trauma* (New York: Penguin, 2015).

Chapter 10: Emotional Paradox

73. **"our emotions are the sources of our most meaningful experiences":** Paul Gilbert, PhD, and Choden, *Mindful Compassion: How the Science of Compassion Can Help You Understand Your Emotions, Live in the Present, and Connect Deeply with Others* (Oakland, CA: New Harbinger Publications, 2014), 59.

74. **"old brain/mind"…is the "base model" of human emotional regulation:** Paul Gilbert, *The Compassionate Mind: A New Approach to Life's Challenges* (Oakland, CA: New Harbinger Publications, 2009), 22.

74. **"new brain/mind":** D. M. Gash and A. S. Deane, "Neuron-Based Heredity and Human Evolution," *Frontiers in Neuroscience* 9 (2015): 209.

75. **negativity bias:** Rick Hanson and Richard Mendius, *Buddha's Brain: The Practical Neuroscience of Happiness, Love & Wisdom* (Oakland, CA: New Harbinger Publications, 2009), 68; Rick Hanson, *Hardwiring Happiness: The New Brain Science of Contentment, Calm, and Confidence* (New York: Harmony Press, 2013).

76. **intentionally compassionate:** Gilbert and Choden, *Mindful Compassion*, 59.

Chapter 11: The Power of a Pause

81. **your empathy networks:** B. C. Berhardt and T. Singer, "Neural Basis of Empathy," *Annual Review of Neuroscience* 35 (2012): 1–23; D. J. Siegel, *Mindsight: The New Science of Personal Transformation* (New York: Bantam Books, 2010), 125.

82. **empathic responses:** T. Singer and O. M. Klimecki, "Empathy and Compassion," *Current Biology* 24 (2014): R875–8.

82. **[mental] training positively affected the brain's neuroplasticity:** B. E. Kok and T. Singer, "Phenomenological Fingerprints of Four Meditations: Differential State Changes in Affect, Mind-Wandering, Meta-Cognition and Interoception Before and After Daily Practice Across Nine Months of Training," *Mindfulness* 8 (1, 2017): 218–231; Kai Kupferschmidt, "Concentrating on Kindness," *Science* 341 (2013): 1336–9, http://science.sciencemag.org/content/341/6152/1336.

82. **by doing these practices, your strengthen your PFC:** J. Brewer et al., "Meditation Experience Is Associated with Differences in Default Mode Network Activity and Connectivity," *Proceeding of the National Academy of Sciences USA* 108 (2011): 20,254–20,259.

82. **the Sacred Pause:** Tara Brach, "The Sacred Pause," *Psychology Today*, December 4, 2014, https://www.psychologytoday.com/blog/finding-true-refuge/201412/the-sacred-pause.

83. **your body posture..., can enhance feeling of compassion:** J. E. Stellar and D. Keltner, "Compassion," in *Handbook of Positive Emotions*, eds. M. M. Tugade, M. N. Shiota, and L. D. Kirby (New York: Guilford Press, 2014), 329–41.

84. **"It is a way to take upon oneself":** Rabbi Rami Shapiro, *The Sacred Art of Loving Kindness: Preparing to Practice* (Vermont: Skylights Path Publishing, 2016) 67.

Chapter 12: Face-to-Heart Connections

88. **our "social nervous system":** Porges, *The Polyvagal Theory.*

89. **compassionate responses activate the vagal nerve and promote well-being:** Dacher Keltner, "The Compassionate Species," *Greater Good Magazine*, July 31, 2012, http://greatergood.berkeley.edu/article/item/the_compassionate_species.

89. **train your brain to be more loving:** Christopher Germer, in personal communication with author, January 5, 2017; adaption from Christopher

Germer and Kristin Neff's curriculum for their eight-week Mindful Self-Compassion Program, Center for Mindful Self-Compassion, https://centerformsc.org.

91. **"compassion isn't some kind of self-improvement project":** Pema Chödrön, *When Things Fall Apart: Heart Advice for Difficult Times* (Boston, MA: Shambhala Publications, 2005), 100 and 104.

91. **"not to be angry or to blame but to try to understand":** Porges, *The Polyvagal Theory*, 75–82; Interview by Ruth Buczynski, "How to Use Polyvagal Theory to Help Patients 'Reset' the Nervous System After Trauma"; Transcript Part I: "Why Polyvagal Theory Holds the Key to Helping Patients Reclaim Safety After Trauma," National Institute for the Clinical Application of Behavioral Medicine.

Chapter 13: Hugs and High Fives

93. **touch is an intimate form of meeting:** David Whyte, *Consolations: The Solace, Nourishment and Underlying Meaning of Everyday Words* (Langley, WA: Many Rivers Press, 2015), 221.

93. **"the energy of tenderness":** Thich Nhat Hanh, *True Love: A Practice for Awakening the Heart* (Boston, MA: Shambhala Publications Inc., 2011).

94. **physical massage can have…a wide range of healing benefits:** Tiffany Field, PhD, *Touch* (Cambridge, MA: The MIT Press, 2014).

95. **pleasant touch…triggers the reward centers in the brain:** I. Gordon et al., "Brain Mechanisms for Processing Affective Touch," *Human Brain Mapping* 34 (4, April 2013): 914–22.

95. **activity in your vagus nerve…elicits feelings of relaxation:** Stellar Keltner, "Compassion," 329–41.

95. **with touch, you get an immune booster:** Field, *Touch.*

95. **oxytocin:** Markus MacGill, "Oxytocin: What Is It and What Does It Do?" *Medical News Today* (September 21, 2015), http://www.medicalnewstoday.com/articles/275795.php?page=2; Taylor et al., "Biobehavioral Responses to Stress in Females: Tend-and-Befriend, Not Fight-or-Flight."

95. **vasopressin supports nurturing and bonding:** Keltner, "The Compassionate Species."

95. **touch communicates emotions:** M. J. Hertenstein, R. Holmes, M. McCollough, and D. Keltner, "The Communication of Emotion via Touch," *Emotion* 9 (4, 2009): 566–73.

95. **skin as our social organ and touch as the social glue:** David J. Linden, *Touch: the Science of the Hand, Heart, and Mind* (New York: Penguin, 2015).

96. **we use [touch] to immediately identify allies or enemies:** Keltner, *Born to Be Good*, 182.

96. **fist bumps, high fives:** M. W. Kraus, C. Huang, and D. Keltner, "Tactile Communication, Cooperation, and Performance: An Ethological Study of the NBA," *Emotion* 10 (2010): 745–9.

96. **rough-and-tumble play:** K. R. Ginsburg et al., "The Importance of Play in Promoting Healthy Child Development and Maintaining Strong Parent-Child Bonds," *Pediatrics* 119 (2007): 182–91.

96. **how necessary is it for well-being, emotional regulation:** Jessica Lahey, "Should Teachers Be Allowed to Touch Students?" *The Atlantic*, January 23, 2015, http://www.theatlantic.com/education/archive/2015/01/the-benefits-of-touch/384706.

97. **"touch seems to be as essential as sunlight":** Diane Ackerman Quotes, https://www.brainyquote.com/quotes/authors/d/diane_ackerman.html.

Chapter 14: Taking in Kindness

101. **"taking in the good":** Hanson, *Hardwiring Happiness*.

Chapter 15: Mindfulness with Heart

107. **mindfulness…is the awareness that emerges through paying attention:** Jon Kabat-Zinn, *Full Catastrophe Living: Using the Wisdom of Your Body and Mind to Face Stress, Pain, and Illness* (New York: Bantam Books, 1990/2013).

108. **training in mindfulness may lead to enduring changes in the brain:** O. Singleton et al., "Change in Brainstem Gray Matter Concentration Following a Mindfulness-Based Intervention Is Correlated with Improvement in Psychological Well-Being," *Frontiers in Human Neuroscience* 8 (2014: 33.).

108. **mindful attention exercises improve emotional regulation, attention:** A. Lutz, A. P. Jha, J. D. Dunne, and C. D. Saron, "Investigating the Phenomenological and Neurocognitive Matrix of Mindfulness-Related Practices from a Neurocognitive Perspective," *American Psychologist* 70 (7, 2015): 632–58.

108. **twenty minutes of mindfulness practice may lead to enduring changes in the brain:** G. Desbordes et al., "Effects of Mindful-Attention and Compassion Meditation Training on Amygdala Response to Emotional Stimuli in an Ordinary, Non-Meditative State," *Frontiers in Human Neuroscience* 6 (2012): 292.

110. **"in the present moment we can learn to see clearly and kindly":** Jack Kornfield, *A Lamp in the Darkness: Illuminating the Path Through Difficult Times* (Boulder, CO: Sounds True, 2014), 10.

111. **"the beautiful mystery of our lives":** Jon Kabat-Zinn, "Jon Kabat-Zinn: Talk and Meditation," presented at Wisdom 2.0, March 2, 2014, https://www.youtube.com/watch?v=Vr2ATJkxzGA.

Chapter 16: Your Loving Self

112. **Dina's story:** Dina Procter, *Madly Chasing Peace: How I Went from Hell to Happy in Nine Minutes a Day* (New York: Morgan James Publishing, 2013); personal interview with the author, January 11, 2017..

113. **"the fraternity of wisdom":** Kornfield, *A Lamp in the Darkness*, 5.

113. **"integration means kindness and compassion":** Dan J. Siegel, *Mind: A Journey to the Heart of Being Human* (New York: W. W. Norton & Company, 2017), 109.

114. **backdraft:** Christopher Germer, The Mindful Path to Self-Compassion, personal interview with the author, January 11, 2017.

114. **the mind needs contrast to know anything:** Christopher Germer, and Kristin Neff, "The Mindful Self-Compassion Training Program," in *Compassion: Bridging Practice and Science: A Multimedia Book*, ed. T. Singer and M. Bolz (Munich: Max Planck Society, 2013), 365–96.

115. **the effects of the MSC program were studied:** Kristin Neff and Christopher Germer, "Being Kind to Yourself: The Science of Self-Compassion," in *Compassion: Bridging Theory and Practice*, 291–312; Kristin Neff and Christopher Germer, "A Pilot Study and Randomized Controlled Trial of the Mindful Self-Compassion Program," *Journal of Clinical Psychology* 69 (2013): 28–44.

115. **self-compassion skills training has been shown to be beneficial:** The scientific study of self-compassion literature can be found at Dr. Kristin Neff's website: http://self-compassion.org/the-research.

Chapter 17: The Naturalness of Being

118. *The Man Who Planted Trees*: YouTube video, 30:07, posted by Pedro Rocha e Mello, October 3, 1987, https://www.youtube.com/watch?v =KTvYh8ar3tc.

119. **The Hidden Life of Trees:** Peter Wohlleben, *The Hidden Life of Trees* (Vancouver, Canada: Greystone Books, 2016), 17.

120. **nature helps to soothe you:** T. H. Kim, et al., "Human Brain Activation in Response to Visual Stimulation with Rural and Urban Scenery Pictures: A Functional Magnetic Resonance Imaging Study," *Science of the Total Environment* 408 (2010): 2600–7; J. Lee et al., "Influence of Forest Therapy on Cardiovascular Relaxation in Young Adults," *Evidence-Based Complementary and Alternative Medicine* (February 10, 2014), http://dx.doi.org/10.1155/2014/834360; L. Tyrväinen et al., "The Influence of Urban Green Environments on Stress Relief Measures: A Field Experiment," *Journal of Environmental Psychology* 38 (2014): 1–9.

120. **being in nature [decreases] rumination:** G. N. Bratman et al., "Nature Experience Reduces Rumination and Subgenual Prefrontal Cortex Activation," *Proceedings of the National Academy of Sciences USA* 112 (28, 2015): 8567–72.

120. **listening to nature...speeds recovery:** E. Sternberg, *Healing Spaces: The Science of Place and Well-Being* (Cambridge, MA: Belknap Press, 2009).

120. **nature...hones attention, inspires creativity, ...spurs kindness:** R. A. Atchley, D. L. Strayer, and P. Atchley, "Creativity in the Wild: Improving Creative Reasoning Through Immersion in Natural Settings," *PLoS One* 7 (2012): e51,474.

121. **psychological benefits of exposure to beautiful nature:** J. W. Zhang et al., "An Occasion for Unselfing: Beautiful Nature Leads to Prosociality," *Journal of Environmental Psychology* 37 (2014): 61–72.

123. **If we surrendered to earth's intelligence:** trans. Anita Barrows and Joanna Macy, R. M. Rilke, *Rilke's Book of Hours: Love Poems to God*, (New York: Riverhead Books, 2005), 171.

Chapter 18: Radical Acceptance

124. **Elyse's story:** Elyse Callahan, "The Do's and Don'ts of Conversing with a Transgender Teen's Mom," *Elephant Journal*, November 9, 2016, http://www.elephantjournal.com/2016/11/the-dos-donts-of-conversing-with-a

-trans-mom/; Elyse Callahan, personal interview with the author, January 13, 2017.

126. **empathy is...a process of deep sensing and listening:** Carl R. Rogers, "Empathic: An Unappreciated Way of Being," *Counseling Psychologist* 5 (1975), 2–10.

126. **"life-alienating communication":** Marshall B. Rosenberg, *Nonviolent Communication: A Language of Life* (Encinitas, CA: PuddleDancer Press, 2015), 15.

Chapter 19: Singing for Our Souls

132. **"hope is the thing with feathers...":** Emily Dickinson, *The Poems of Emily Dickinson*, ed. R. W. Franklin (Boston: Harvard University Press, 1999).

132. **"laughter, song, and dance create emotional and spiritual connection":** Brené Brown, *The Gifts of Imperfection: Let Go of Who You Think You're Supposed to Be and Embrace Who You Are* (Center City, MN: Hazelden Publishing, 2010), 118.

132. **"co-pathy":** S. Koelsch et al., "Investigating Emotion with Music: An fMRI Study," *Human Brain Mapping* 27 (2006): 239–250; S. Koelsch, "Brain Correlates of Music-Evoked Emotions," *Nature Reviews Neuroscience* 15 (2014): 170–80.

133. **your experiences of music and singing have deep neurological correlates:** Porges, *The Polyvagal Theory.*

133. **music is simply good for your health:** M. Chanda and D. J. Levitin, "The Neurochemistry of Music," *Trends in Cognitive Sciences* 17 (4, 2013): 179–93; S. Garrido et al., "Music and Trauma: The Relationship Between Music, Personality, and Coping Style," *Frontiers in Psychology* 6 (2015): 977; T. Eerola, J. K. Vuoskoski, and H. Kautiainen, "Being Moved by Unfamiliar Sad Music Is Associated with High Empathy," *Frontiers in Psychology* 7 (2016): 1176; L. Taruffi and S. Koelsch, "The Paradox of Music-Evoked Sadness: An Online Survey," *PLoS One* 9 (2014): e110490; V. N. Salimpoor et al., "The Rewarding Aspects of Music Listening Are Related to Degree of Emotional Arousal," *PLoS One* 4 (2009): e7487.

133. **empathy and sad music:** A. Kawakami and K. Katahira, "Influence of Trait Empathy on the Emotion Evoked by Sad Music and on the Preference for It," *Frontiers in Psychology* 6 (2015): 1541; T. Rabinowitch, I. Cross, and P. Burnard, "Long-Term Musical Group Interaction Has a Positive Influence on Empathy in Children," *Psychology of Music* 41 (2013): 484–98.

134. **music...will have social benefits:** D. Levitin, *This Is Your Brain on Music: The Science of a Human Obsession* (New York: Plume, Penguin Group, 2007); K. Overy and I. Molnar-Szakacs, "Being Together in Time: Musical Experience and the Mirror Neuron System," *Music Perception: An Interdisciplinary Journal* 26 (2009): 489–504; K. M. Kniffin, J. Yan, B. Wansink, and W. D. Schulze, "The Sound of Cooperation: Musical Influences on Cooperative Behavior," *Journal of Organizational Behavior* 38 (2017): 372–90.
134. **synchronized musical activities:** Girish, *Music and Mantras: The Yoga of Mindful Singing for Health, Happiness, Peace and Prosperity* (New York: Enliven Books, 2016), xviii.
134. **"love, I find, is like singing":** Zora Neale Hurston, *Dust Tracks on a Road* (New York: HarperCollins, 1942), 203.

Chapter 20: Cherishing the Little Things

138. **Be loving. Show it. Say it:** Cheryl Strayed, *Tiny Beautiful Things: Advice on Love and Life from Dear Sugar* (New York: Random House, 2012), 18.
138. **"I love you" should not be limited:** Andy Gonzalez, Holistic Life Foundation, personal interview with co-founder, September 21, 2016.
138. **"positivity resonance":** Fredrickson, "Positive Emotions Broaden and Build," 40–3.
139. **the secret to stable and lasting relationships:** John M. Gottman and Nan Silver, *The Seven Principles for Making Marriage Work: A Practical Gruide from the Country's Foremost Relationship Expert* (New York: Harmony, 1999/2015), 21.
140. **the secret to love is just kindness:** Emily Esfahani Smith, "Masters of Love: Science Says Lasting Relationships Come Down to—You Guessed It—Kindness and Generosity," *The Atlantic*, June 12, 2014, http://www.theatlantic.com/health/archive/2014/06/happily-ever-after/372573/.
141. **master couples are healthier and live longer:** J. M. Gottman, "What Do Women Really Want?" The Gottman Institute, January 18, 2016, https://www.gottman.com/blog/what-do-women-really-want/.
141. **the "cherishing" skill:** Gottman and Silver, *The Seven Principles for Making Marriage Work.*

Chapter 21: An Attitude of Gratitude

143. **express sentiments of appreciation:** Erika Lantz, producer, *Kind World*, WBUR, Boston, personal interview with author, March 3, 2017.

145. **"I recognize, I acknowledge, I am grateful":** D. Steindl-Rast, "A Deep Bow," *Main Currents in Modern Thought* 23 (1967): 129–32, http://gratefulness.org/resource/a-deep-bow/.

145. **gratitude is an inclination toward noticing and appreciating the positives in life:** A. M. Wood, J. J. Froh, and A. W. A. Geraghty, "Gratitude and Well-being: A Review and Theoretical Integration," *Clinical Psychology Review* 30 (2010): 890–905.

145. **the evolutionary underpinnings of gratitude and reciprocal altruism:** M. Suchak, "The Evolution of Gratitude," *Greater Good Magazine*, February 1, 2017, http://greatergood.berkeley.edu/article/item/the_evolution_of_gratitude.

145. **"the moral memory of mankind":** *Greater Good Magazine*, "What Is Gratitude?" http://ggsc-web-prod-01.ist.berkeley.edu/topic/gratitude/definition#what_is.

146. **the psychological, physical, and social benefits of gratitude:** Wood, Froh, and Geraghty, "Gratitude and Well-being; L. S. Redwine et al. "Pilot Randomized Study of Gratitude Journaling Intervention on Heart Rate Variability and Inflammatory Biomarkers in Patients with Stage B Heart Failure," *Psychosomatic Medicine* 78 (6, 2016): 667–76.

146. **four ways gratitude contributes to our lives:** R. Emmons, "Why Gratitude Is Good," *Greater Good Magazine*, November 16, 2010, http://greatergood.berkeley.edu/article/item/why_gratitude_is_good.

146. **the difference between indebtedness and gratitude:** M. A. Matthews and J. D. Green, "Looking at Me, Appreciating You: Self-Focused Attention Distinguishes Between Gratitude and Indebtedness," *Journal of Cognition and Emotion* 24 (2010): 4, http://www.tandfonline.com/doi/abs/10.1080/02699930802650796.

147. **benefits in mutual nurturing:** M. A. Mathews and N. J. Shook, "Promoting or Preventing Thanks: Regulatory Focus and Its Effect on Gratitude and Indebtedness," *Journal of Research in Personality* 47 (2013): 191–5.

147. **"taking in the good":** Hanson, *Hardwiring Happiness*.

Chapter 22: Grace and Grit

152. **kindsight "what a liberating thing":** Krista Tippett, "Reconnecting with Compassion," TED Talk, February 2011, https://www.ted.com/talks/krista_tippett_reconnecting_with_compassion/transcript?language=en.

153. **positive emotions…broaden and build inner your resources:** Fredrickson, *Positivity*.

153. **studies from the Positive Emotions and Psychophysiology Laboratory:** B. L. Fredrickson et al., "Open Hearts Build Lives: Positive Emotions, Induced Through Loving-Kindness Meditation, Build Consequential Personal Resources," *Journal of Personality and Social Psychology* 95 (5, 2008): 1045–62; Fredrickson, "Positive Emotions Broaden and Build," 1–53.

154. **an ability to "savor future events":** Fredrickson, "Positive Emotions Broaden and Build," 1–53.

154. **"bounce back" from stressful or negative experiences:** M. M. Tugade and B. L. Fredrickson, "Resilient Individuals Use Positive Emotions to Bounce Back from Negative Emotional Experiences," *Journal of Personality and Social Psychology* 86 (2, 2004): 320–33.

154. **"grit":** Angela Duckworth, *Grit: The Power of Passion and Perseverance* (New York: Scribner, 2016).

Chapter 23: Forgiveness Is a Gift to Yourself

158. **"if you want to find a way out, go in":** Byron Katie, "Inquiry, Politics & the Mind," presentation given at Wisdom 2.0, San Francisco, CA, February 18, 2017, http://wisdom2conference.com/Speakers.

159. **people who have been able to forgive…enjoy:** G. Bono, M. E. McCullough, and L. M. Root, "Forgiveness, Feeling Connected to Others, and Well-Being: Two Longitudinal Studies," *Personality and Social Psychology Bulletin* 34 (2008): 182–195; A. H. Harris et al., "Effects of a Group Forgiveness Intervention on Forgiveness, Perceived Stress, and Trait-Anger," *Journal of Clinical Psychology* 62 (6, 2006): 715–33.

160. **forgiveness is "the experience of peace or understanding that can be felt in the present moment":** Fred Luskin, *Forgive for Good: A Proven Prescription for Health and Happiness* (New York: HarperCollins, 2002).

160. **"untie the knot" and "release your prisoners":** Janet Conner, *The Lotus and the Lily: Access the Wisdom of Buddha and Jesus to Nourish Your Beautiful, Abundant Life* (San Francisco, CA: Conari Press, 2012), 107–15;

Janet Conner, "Releasing Your Prisoners Meditation," http://janetconner.com/manifestabundantlife.

162. **"forgivishness":** Anne Lamott, *Grace (Eventually): Thoughts on Faith* (New York: Riverhead Books, 2008), 145.

Chapter 24: Apologies Make for a Friendly World

165. **three important components to apologies:** Ryan Fehr and Michele J. Gelfand, "When Apologies Work: How Matching Apology Components to Victims' Self-Construals Facilitates Forgiveness," *Organizational Behavior and Human Decision Process* 113 (2010): 37–50, http://www.gelfand.umd.edu/fehr_gelfand_obhdp.pdf.

166. **three perspectives to consider [when accepting an apology]:** Fehr and Gelfand, "When Apologies Work."

166. **understanding a victim's values and needs:** Aaron Lazare, MD, "Making Peace Through Apology," *Greater Good Magazine*, September 1, 2004, http://greatergood.berkeley.edu/article/item/making_peace_through_apology; Aaron Lazare, MD, "What an Apology Must Do," *Greater Good Magazine*, Accessed September 1, 2004, http://greatergood.berkeley.edu/article/item/what_an_apology_must_do/.

Chapter 25: Writing New Beginnings

172. **"we tell ourselves stories in order to live":** Joan Didion, *The White Album* (New York: Farrar, Straus and Giroux, 1979/2009).

172. **the physical and emotional benefits of expressive writing:** James W. Pennebaker and J. M. Smyth, *Opening Up by Writing It Down: How Expressive Writing Improves Health and Eases Emotional Pain* (New York: Guilford Press, 2016).

173. **"morning pages":** Julia Cameron, *The Artist's Way: A Spiritual Path to High Creativity* (New York: Putnam, 2002).

173. **"rumbling":** Brené Brown, *Rising Strong: The Reckoning. The Rumble. The* Revolution. (New York: Spiegel & Grau, 2015), 91.

173. **sharing what you write:** Brené Brown, "Finding Shelter in a Shame Storm (And Avoiding the Flying Debris)," Oprah.com, 2013, http://www.oprah.com/spirit/brene-brown-how-to-conquer-shame-friends-who-matter#ixzz4ZpcWC0MA.

174. **reflect with compassion:** James W. Pennebaker, *Writing to Heal: A Guided Journal for Recovering from Trauma and Emotional Upheaval.* (Oakland,

CA: New Harbinger Publications, 2004); James W. Pennebaker, "Writing and Health," https://liberalarts.utexas.edu/psychology/faculty/pennebak #writing-health.

174. **your future self:** Cameron, *The Artist's Way*, 89.

174. **watch a poem unfold:** Mary Pipher, *Writing to Change the World: An Inspiring Guide for Transforming the World with Words* (New York: Riverhead Books, 2006), 32.

Chapter 26: Accepting a Helping Hand

178. **"weave a web of connection with the world":** Anaïs Nin quotes, https://www.goodreads.com/quotes/612430-all-of-my-creation-is-an-effort-to-weave-a.

178. **social connection is broadly defined:** E. Seppala, T. Rossomando, and J. R. Doty, "Social Connection and Compassion: Important Predictors of Health and Well-being," *Social Research* 80 (2, 2013): 411–30.

178. **one in four Americans:** M. McPherson, L. Smith-Lovin, and M. E. Brashears, "Social Isolation in America: Changes in Core Discussion Networks over Two Decades," *American Sociological Review* 71 (2006): 353–75.

178. **living alone and feeling lonely are risk factors for mortality:** J. Holt-Lunstad et al., "Loneliness and Social Isolation as Risk Factors for Mortality: A Meta-Analytic Review," *Perspectives on Psychological Science* 10 (2, 2015): 227–237.

179. **social connectedness...and well-being:** Christakis and Fowler, "Friendship and Natural Selection."; E. Seppala, T. Rossomando, and J. R. Doty, "Social Connection and Compassion: Important Predictors of Health and Well-being," *Social Research* 80 (2, 2013): 411–30.

179. **vocalizations...are adaptive for human bonding:** Keltner, *Born to Be Good*.

179. **social connections [trigger] the neural reward circuitry:** U. Wagner et al., "Beautiful Friendship: Social Sharing of Emotions Improves Subjective Feeling and Actives the Neural Circuitry," *Social and Affective Neuroscience* 10 (6, 2015): 801–8.

181. **giving is the great equalizer:** Stephen G. Post and Jill Neimark. *Why Good Things Happen to Good People: How to Live a Longer, Healthier, Happier Life by the Simple Act of Giving* (New York, NY: Broadway Books), 6.

Chapter 27: Gifts in Kind

184. **"loving-kindness and compassion are the two faces of altruism":** Ricard, *Altruism*, 25–6.
185. **"altruists see a fellow human being":** Kristen Renwick Monroe, *The Heart of Altruism: Perceptions of a Common Humanity* (Princeton: Princeton University Press, 1996).
185. **"paying it forward":** M. Tsvetkova and M. Macy, "The Science of 'Paying It Forward,'" *The New York Times Sunday Review*, March 14, 2014, https://www.nytimes.com/2014/03/16/opinion/sunday/the-science-of-paying-it-forward.html; *Greater Good Magazine*, "What Is Altruism?" http://greatergood.berkeley.edu/topic/altruism/definition.
185. **global survey of empathy levels:** W. J. Chopik, E. O'Brien, and S. H. Konrath, "Differences in Empathic Concern and Perspective Taking Across 63 Countries," *Journal of Cross-Cultural Psychology* 48 (1, 2016): 23–38.
186. **review quality studies on kindness interventions:** O. S. Curry et al., "Happy to Help? A Systematic Review and Meta-Analysis of the Effects of Performing Acts of Kindness on the Well-being of the Actor," *Perspectives on Psychological Science*, September 21, 2016, https://osf.io/ytj5s.
189. **[generosity] "frees up oceans of energy":** Lynne Twist, *The Soul of Money: Transforming Your Relationship with Money and Life* (New York: W. W. Norton & Company, 2003).

Chapter 28: Networks of Generosity

191. **"we are all part of humanity…":** C. Kelly-Gangi (ed.), *The Dalai Lama: His Essential Wisdom* (New York: Fall River Press, 2007), 36.
192. **social norms…human social networks:** James H. Fowler and Nicholas A. Christakis, "Cooperative Behavior Cascades in Human Social Networks," *Proceedings of the National Academy of Sciences USA* 107 (12, 2010): 5334–8; J. Surma, "Social Exchange in Online Social Networks: The Reciprocity Phenomenon on Facebook," *Computer Communications* 73 (Part B, 2016): 342–346.
193. **Three Degrees of Influence Rule:** N. A. Christakis and J. H. Fowler, *Connected: The Surprising Power of Our Social Networks and How They Shape Our Lives* (Boston: Back Bay Books, 2011).
193. **"we form social networks because…":** Nicholas Christakis, "How Do Our Networks Affect Our Health?" in interview with Guy Raz, *TED Radio Hour*, March 4, 2016, http://www.npr.org/templates/transcript/transcript.php?storyId=468881321.

Conclusion: Reimagine Kindness

198. **"failures of kindness":** Joel Lovell, "George Saunders's Advice to Graduates," *New York Times*, July 31, 2013, https://6thfloor.blogs.nytimes.com/2013/07/31/george-saunderss-advice-to-graduates/?_r=0.

199. **"love and resist at the same time?"** Sharon Salzberg, "To Love or Resist: Why Not Both?" presentation given at Wisdom 2.0, San Francisco, February 18, 2017.

Tara Cousineau, PhD, is a clinical psychologist, meditation teacher, well-being researcher, and social entrepreneur. She has received numerous grants from the National Institutes of Health Small Business Innovative Research program. Cousineau founded www.bodimojo.com, and develops digital wellness tools for youth. She is affiliated with the Center for Mindfulness and Compassion at Cambridge Health Alliance in Somerville, MA. She is mindfulness trainer and chief science officer at Whil, a digital mindfulness company, and serves as a scientific advisor to www.kindness.org. She is dedicated to global efforts to spread kindness in both small ways and large.

Foreword writer **Stephen Post, PhD**, is coauthor of the bestseller, *Why Good Things Happen to Good People*. He has been quoted in *The New York Times*, *Parade*, *O, The Oprah Magazine*, *U.S. News & World Report*, and *Psychology Today*. He has also been interviewed on numerous television shows, including *Nightline*, *The Daily Show*, and *Stossel*. A transformative speaker and thought leader, Post has inspired thousands with medical and philosophical knowledge based on thirty years of research.